AF375441

RED CHILD

RED CHILD

Jess Donoho

JessDonoho.com

RED CHILD

Jess Donoho

JessDonoho.com

Cover photo courtesy of BRANO (unsplash.com)

This is a work of fiction. I don't generally set out to write a specific book. I write about things that fascinate me and one thing leads to another, and soon a story is unfolding. I do not plot out the course of the book with a beginning, middle and ending. I don't know what will happen until I write it, so I am just as excited about what is happening next as the reader is. My books are never formula. There is rarely a love interest, a villain or a hero. Instead, there are characters that unfold as the story does. I find this to be less predictable, and more fun to experience.

You are very likely to find typographical and grammatical errors in my books. I do not have editors, proofreaders', agents, or publishers. I love the process of producing the entire story myself. If you want perfection, there are a million authors with master's degrees writing predictable literature. If you enjoy a good story, I am your storyteller. I hope you find happiness in the RED CHILD. Please let me know what you think at jessdonoho.com.

Dedicated to my children who are the source of my happiness, and my mother, who is the strongest woman I know. - Jess

CONTENTS

CONTENTS

CONSUMATION

The infant was born as all infants are, in the moment of passion and intimacy. When two come together to consummate their pairing deep in the core, below the chromosphere, the tachycline and into the radiative zone. Here, in the planet-melting center of the galaxy, life is being created in a blistering cauldron of gas and fire. As their energies unite, the landscape ripples with heat. These ripples give way to waves, which become tsunamis of lava. Tornados of solar winds form spectacular prominences. As the two reach their apex, fountains of molten iron and gas are spewed miles into the air. Great cyclones of fire and a blinding incandescence increases until entire planets millions of miles away are seared and baked. The entire galaxy of stars and planets is engulfed in this romantic gesture, and then it subsides. The violence will continue on the surface for an eternity, but under the surface, deep in the core lies a cluster of eggs that will incubate until they hatch, giving birth to the Children of the Sun.

During the consummation, the fiery blast completely engulfs Mercury, crystallizing the surface into a glass orb. The exposed surface of Venus is enveloped in fire and its seas dry up and forests and mountains are turned to dust, leaving behind a dead hulk that will require re-seeding and re-growing over millions of years. Earth was hit with a decreasing amount of the Suns fierce energy. Although it suffered greatly on the side facing the fiery star, the portion in the Suns shadow fared better. Life would resume with great variety. The solar storm fanned out across the galaxy. Mars, Jupiter, Saturn, and Neptune all feel decreasing effects the farther from the Sun they spin, forcing all life deep underground for millennia. Tiny Pluto, on the outskirts of the solar system, barely felt a whisper. Wave after wave continued outward into other galaxies as a ripple in space, proclaiming that in this one small corner of the Universe, a spectacular event had occurred.

As is the course of nature in our System, life throughout the galaxy is swept aside, then it is rebirthed again and again. This is a story of the Sun, but also of its third planet, Earth. The two are interconnected, and as our story threads and winds its way through eons, their fates are tied. It is in this cycle of birth and death that a single creature rises from the depths of Earth's surface, through muck and mire, from fin to scale to wing and foot. Its evolutionary path a drunkard's stagger through time. A creature whose form is entirely flawed, but whose intelligence allows it to submit all other creatures on this sphere. This is the story of the rise of humankind. A story of cunning, fearlessness and evil that all humans embody, even encourage in their civilizations.

†††

Mer and Zel settled into one another. Their bodies rippling with the heat of post-coital bliss. Their fingers intertwined and their eyes closed to the light. They were nestled in a gaseous stew of hydrogen and helium, but breathed deep of the oxygen, that made up a small portion of its atmosphere. As the procreator's, they alone created life on this star. Each clutch of eggs produced would slowly build the population of this massive place. Each child of the sun would add lives to a star that made up nearly 99% of the total mass of the solar system. Each consummation of their love would cause massive upheaval and storm throughout the system, forever changing the landscapes and lives of those living there.

Children of the Sun would be born as innocents, having neither instinct nor past life memory to guide them. They would learn as they lived and be shaped by their environment and the society of others that populated the sun. As a gaseous star, the sun lacks the topography and geography of solid planets, but aside from that, societies form and tear like societies throughout the universe. Alliances are made as are enemies. Rather than war, there is simply submission, with the highest intelligence and intensity always winning out over brute force. In this way, it is far more civilized than the solid planets, whose inhabitants fight for superiority and eventually succumb to extinction, only to be replaced by another species that climbs the evolutionary ladder to dominance. The cycle is never ending and plays out over millions of years as the universe watches the theater unfold.

Within the sun's society, the Creator's had a unique place, for they were the sole source of procreation. As communities and factions of the sun sought to dominate others, all respected and bowed before the Creator's. To eliminate them would be to doom all. Beyond that, the Creator's did not participate in the politic and play of the Children. They did not have opinions or subject others to their authority. They simply represented undying love and the example of what true love and happiness was. No matter the conflict within the communities, all aspired to their example. Leading by example is the Universal foundation of all learning.

Mer and Zel floated in the gaseous photosphere, the coolest layer of the Sun. As they drifted, they surveyed their Children, encouraging them and offering gifts of carbon and cobalt, which were coveted as both prize and power in this place. The Sun has little to offer in elemental reward. Cobalt, carbon, and iron are elements that are prized above all others. They are a solid feature in a gaseous world. To own these elements is to adorn yourself with them and show it off. To have mined it from the star itself is legendary, but to have it bestowed upon you by the Creator's themselves is a rare honor.

On a Star the size of the Sun, it takes many thousands of years for the Creator's to completely traverse the diameter of the globe and millions of solar years to cover its entire surface. In the history of the sun, only a handful of Children of the Sun have received more than one gift of carbon or iron. These are known as the Legion. They are the wisest and most compassionate of Children of the Sun. They are icons and examples of what all Children should be. They serve to guide and lead. Their gifts are draped over their head in a net-like covering whose black or red fibers contrast with the irides-cent glow of their bodies.

There are even fewer Children who have received two gifts in their lifetimes. These are called the Holy, and they mediate and deescalate disagreements and conflict. Their net-like gifts form a cowl over their head and drape over their shoulders, forming a short cape of black and brown.

There has only been one Child who has been completely gilded with gifts. His cowl is a solid sheet of carbon and iron, completely covering his head and draping down his body in layers. Elemental neon is woven into the garment, which sparks and glows in the intense solar heat. It ripples with the solar winds. This is the Lord. The judge, jury, and executioner of justice. The Lord is not an example of peace and happiness. He is a terrifying reminder of consequence. To be visited by the Lord is not a gift, it is a penalty, which is carried out in ways that are unspeakable and unthinkable. Every society needs a reminder of true consequence for your words and actions. The Lord embodies this to his very soul. To the credit of the sun's society, his judgement was only necessary every few centuries.

The Lord is a curiosity, for while the reader might equate his task to be an aggressor or heavy handed, he was neither. The Lords' nature was not embodied in what the planet-dwellers would call evil. He was an aberration, made by the Creator's specifically for this task. No matter how perfect your children are, there will always be those few that are broken. Those that would seize rather than earn. Who would cheat rather than build. Those who did not learn by example or looked only for the lines between goodness that would benefit them. These needed a consequence and the Lord exemplified consequence in a most horrific way.

The Lord lived in the sun's core, in the most inhospitable place on an inhospitable star. He simmered in a pool of heat and energy. He did not wear his gifts like a crown, but like shackles. He despised his gifts. He resented them, but he wore them of obligation. Such was his embodiment of sympathy, empathy, and care, that only he could wield the gifts of the Lord. If he were of less spirit, the gifts would consume him entirely. He wore the full weight of these gifts, wrapped around his body. He executed the will of the gifts as was required, fighting every cell of his bodily conciseness to maintain his role. His gift was a curse, and this knowledge insured that all Children of the Sun honored and respected the Lord, even as they feared him. Such great sacrifice in service to the star. Loved by all, feared by all.

The greatest punishment any Child of the Sun could endure was not death, but to be sentenced as the servant of the Lord. These children were named the Damned and were forever bound to the Lord. They were the hunters of those who would violate law. Each Damned is a witness to the Lord's judgement and execution. The occurrence of judgments was rare in this fiery world, but each was far more painful to witness than to endure, for the punishment was less a reminder to the offender and more a lesson to all of consequence. Eventually, the wounds of punishment would heal, but the memory of the execution would remain in the hearts and minds of all who witnessed it forever.

†††

The only blessing of the gift was that the Lord was also the keeper of the clutch. The eggs that would one day be the new residents of this star. As the Lord walked the halls that were encrusted with eggs, he spoke to them. He read poetry; he spoke of good and evil and of justice and community. As they incubated, their inner light radiated outward, bathing the corridors in an ethereal blue light. Not all eggs were the same size or shape. Each was unique, as was the precious life inside them. Whether smoothly round or sharply angular, they all burned blue with the warmth of the love of the Creator's. All except for one. In the many clutches of eggs that the Lord had nurtured to birth, this was the anomaly. It did not burn blue and clear, but was a milky red. The gasses inside the egg did not gently swirl as with the others, the gasses were a tumultuous storm that never relented. As the Lord walked past the egg daily, he often thought of destroying it before it could hatch. It reeked of negative energy. It smelled like death itself.

As the year of hatch approached, the Lord summoned his Damned and sent them ahead to announce his coming. He left his home in the core and glided through the layers of the Star to the surface. As he came into the thermocline, where most of the Children resided, they gathered in large crowds to witness his coming. Many had never seen the Lord, only heard the tales of his existence. There was no mistaking the Lord from the others. His cloak was a bright neon sign proclaiming his coming. With the hood obscuring his face, there was no emotion conveyed, which was frightening to a people who rely greatly on facial emotion to communicate.

Still, there was a cheer and salute as the Lord passed their way. They stared in terrified awe, hoping he was not coming for them. As he passed, their relief was palpable. They had seen the Lord, and they were safe. This would be the topic of discussion for years to come, 'the day the Lord passed by.'

Mer and Zel were prepared for the Lord as he came into the photosphere. Gifts of oxygen and neon were given. The neon setting over the form of the Lord as he breathed in the precious oxygen. So rare was oxygen that it was a most terrific enhancer of physical being. An immediate sense of alertness and confidence hit you as you inhaled its fragrance. It was a gift the Lord welcomed and appreciated. The Lord then bowed low to the Creator's, and they all reclined on a thermal updraft of solar wind.

Zel was first to speak with a formal welcome to the Lord. She called him Lord Ras, as this was both title and his Child name. He closed his eyes and nodded deeply in thanks for the welcome. Mer then asked the reason for this untimely visit.

"There is an unnatural occurrence in the egg clutch," said the Lord. He went on to explain the phenomenon of the red egg and his concerns.

In all the eons of time, this was truly an irregularity and it greatly concerned the Creator's. Questions were asked about shape and size. Of placement and location in the clutch. Nothing more than the color and physical attributes was out of the ordinary. The Lord offered his opinion that it might be wise to destroy the egg now rather than allow it to hatch. Both Mer and Zel were stunned by this opinion. Never had an innocent been judged and executed without the opportunity to prove itself. "In fact, perhaps this egg was special in a positive and healthy way," offered Zel. "Shouldn't we wait to see what it will become before we doom it too death?"

The Lord did not answer, he had offered his opinions and now it was his task to execute the will of the Creator's. He sat quietly while they debated the practical and the emotional consequences of actions taken or not. In the end, as it is throughout the Universe, the feminine won out and Zel turned to the Lord and said, "if it hatches, it shall be borne to the Damned. You yourself shall raise it as your own. If it is an offense to us, you shall judge and execute at your discretion."

This was not at all pleasing to the Lord. He was shackled to these Gifts and now shackled to a hatchling. Moreover, he would likely be the death of this Child. The Lord thought for a long moment, then bowed low, turned, and drifted downward.

With his Damned announcing his coming, the throngs of Children in an entirely different part of the spinning star gathered to witness the descent of the Lord. Although they could see no emotion on his masked face, the foul disposition of the Lord was evident. There was no cheer, no salute, only a terrified, shrinking back of the crowd, who quickly dispersed into the ether lest they attract his attention. When one of such power was in a foul mood, you do not stick around to gawk at his presence. The Lord descended further into the depth and layers of the star until he reached the core. He walked directly to the bloodied egg, which he removed from the clutch and tucked into the folds of his cloak. There is a universal saying to keep your friends close and your enemies closer. the Lord was not taking any chances.

The hatching was the most cheerful day in the Lords otherwise grim existence. It begins as a tapping sound. With the first tap, the Lord is on his feet scrambling up and down the corridors looking for the first hatchling. When he finds it, there is already a crack appearing in the shell. Soon, a flame darts from the thin space in the smooth oval. Others are beginning to tap and move, but the first holds his attention. Fingers reach through the crack, grasp at shell, and violently begin to pull and claw it apart. It is as if the occupant is desperately trying to escape. The sharp edges of the shell cut at the skin of the fingers and hand, leaving bloody blue streaks across the translucent surface of the shell. When the fracture is wide enough, a head emerges, gasping and crying out for freedom. The rage intensifies. The Child is frantic to escape this prison that has held it for centuries as it awaited hatching. When it finally emerges, its solid form falls to the floor of the corridor as it pants in exhaustion. Within minutes, its solid form begins to lightly vaporize at the surface. This vaporization creates a crust layer of skin which in moments oxidizes before turning to a scaly dust across the form. It is a molting of egg-skin and the birthing of new-skin.

Other Children are flailing their arms as they seek to escape their egg. Bodies are falling out of their eggs on to the floor, crusting and oxidizing. Hundreds, then thousands of them are hatching and meta-morphing as the Lord watches. He sings to them as they erupt from their shells, pacifying their anguish and fear. His song tells them that they are welcome to this new life. The Lord is overcome with love and happiness for this moment. His cloak is no longer a prison. His execution is set aside to welcome the new hatchlings. His children, the Children of Mer and Zel, the Children of the Sun.

The Lord glided above the forms of the sleeping Children in the corridors. Broken shell was being collected by the Damned and the walls scraped of any evidence of the nursery. As he glided, the Lord sang softly to his Children. He sang of the history of the Sun, of the solar system and of the cosmos. He sang of the blessings of life and the terrible anguish of judgment and execution. He sang of his desire to never visit them as executioner, that they should be kind citizens. That they should be respectful and care for each other. That their lives and their choices were theirs alone. They could choose a life of happiness or a life of pain. He sang of his well wishes for their choice of happiness.

One by one, the Children woke. They stretched and yawned. They flexed their new limbs and wiggled their new toes. As they began to stand, they lifted off the ground and floated down the corridors to the vents that would lead them to the surface. Like a torrent, they shot upwards though the vents by the tens of thousands, erupting into the thermocline to the cheers of the those who had been waiting for the new hatch. As they settled to the surface, each was greeted with embraces and words of encouragement. They were home.

When the last of the Children was gone, and the final scraps of shell and skin was cleared from the corridors, the Lord cheerfully returned to his place in the Core. After the excitement and noise of the hatch, the quiet was deafening and peaceful. The Lord closed his eyes and drifted into a sleep state of peace. For the first time in centuries, he was smiling and cheerful. His heart was full, and his mind was a peace. He pulled his cloak closer to his body, swaddling himself in a cocoon of comfort, only to feel the lump of egg in the cloak that was the Red. In his excitement he had forgotten the red egg. With a furrowed brow he removed it from the folds of his cloak and observed it in the incandescent glow of the suns core. Nothing had changed. It was the same milky-red swirling center that had existed for all these years. No tapping, no cracking. Worried, he returned the egg to his cloak and all but forgot it for decades.

†††

It was a grim time of tracking. The Damned were spread out through the tachycline, crossing over into the radiant zone, looking for a Child who had been judged and convicted to become one of the Damned. He had fled and was moving across the Star transferring between his physical self and his vapor self to avoid detection. The Damned were not deterred, only slowed by this action. This was his first attempt at fleeing, but the Damned had been tracking this type of evasion for centuries. They were patient, as was their

Lord. As the Lord floated high above the tachycline, he sensed rather than heard a tapping. It was a familiar sound, but one that he could not place. It was not until he felt the fingers tearing at the egg in his robe did he realize that his red egg was indeed hatching. He sent news to the Damned to bring the convict to him when he was caught, and he quickly descended into the core. He floated rather than walked down the corridor to the very place that he had removed the red egg. He gently removed it from his pocket and lay it in the corridor. He then squatted low and watched it fight for release. The fingers were far smaller than a normal Childs, but they were strong. Where a normal Child of the Sun would tear small pieces of shell out fist by fist, the red egg was coming apart in large sections. In short time, a face appeared in the shells break. It was not frantically gasping and fighting for release, there was a look of intense and deliberate effort. A determination and sense of occupation. This Child was dismantling an obstacle, not fighting free. When it at last emerged from the egg it stood on wobbly legs and shuddered. As its body shook, the skin turned to a scabrous crust. The Red Child fought back pain and weakness and tightened every muscle in its sinewy and thin body. It remained in this tense state for long minutes as the scabs hardened. At last, the crust fractured and dropped to the floor in dust. The Red Child remained standing. She was panting and weak from the effort, but she never dropped to the floor. She never cried out and she never allowed the change to own her. As the Lord watched, the Child raised her large eyes and look straight at the Lord, no, she looked straight *through* the Lord.

She had no fear, no respect and no sense of awe or wonder. At birth, this Child was the Lord's superior, and she sensed it. A fear ran through the Lord. His intuition had told him to destroy the egg decades ago and now it had been allowed to hatch. This was the beginning of something new, and dangerous.

THE RED CHILD

The Red Child took her place among the Damned. She never spoke, she always followed instructions and she always moved with a grace and efficiency that was unmatched by any other. She was a runt, barely one-third the size of most Children of the Sun. She seemed not to notice as she took on increasing responsibilities. There is no rank among the Damned. They are all simple servants conducting a task, yet the Red Child had a confidence and purpose that demanded attention. When she moved, the Damned moved with her. Where she went, they followed and protected. She was naturally the leader and the Damned sensed it. Within this informal rank was the Lord, who watched the Red Child carefully. While the Red Child was obedient, one had the feeling that at any moment she could end your existence without thought or remorse.

The life span of a Child of the Sun is immeasurable. That is to say that no Child pays attention to age or is concerned by it. Unless the Lord ends your existence, it seems to go on for an eternity. Centuries passed with the Lord serving as the master of the Red Child. Neither showed the other any affection, yet there was a bond there. A mutual respect. For although the Lord was far more ancient, the Red Child was his equal from the day she stepped onto the sun's surface. After time, it seemed that they conducted the Damned in concert, with the Lord issuing the order of execution and the Red Child bringing the accused to judgment before the Lord. It was inevitable that one day the Red Child would issue the execution herself. The hunt was on, the prey was caught and the Red Child, without a second thought, judged and ended the existence of the accused. The Damned were horrorstruck. This was the sole domain of the Lord. For any of the Damned to think of assuming this command was unspeakable.

The Lord came upon the scene in a rush. His fury and anger creating a tornado of flames. Explosive gasses whipped across the Damned who scattered and hid. Tendrils of fire enveloped the Red Child, who stood as passive and unmoving as always. Without a word, the Lord enveloped the Red Child in a cocoon of molten gasses and began the ascent to the photosphere. He lashed out at the Damned to hurry ahead to warn the Children and Mer and Zel of his coming, and rage.

The Red Child tried to tear at the gasses that bound her but found that she did not have the strength of power to achieve it. She reached deep into her spirit to shoot her own flames outward, but the binds were too tight. For hours she fought without rest. For hours they ascended. When they reached the tachycline, there were no Children there to greet them. They had heard of the Lord's wrath and the unspeakable violation of the Red Child. This was no time to invite the Lord's wrath upon them. As the Lord ascended through the empty cities and into an empty red sky, the Damned fell into formation behind him. Up into the photosphere, the coolest layer of the sun's atmosphere, where Mer and Zel were awaiting his arrival. They were flanked on either side by the Holy, and behind them the Legion. The two groups facing each other as if meeting on a field of battle.

The Lord floated the Red Child to the space between the two sides. As he went, he unraveled the cocoon from the head of the Red Child, leaving the wrappings around her body. Mer and Zel had never witnessed this level of ferocity from the Lord. This was a new experience for them all and demanded absolute caution. They did not ask the Lord for an explanation, nor did they ask it from the Red Child. They had never seen this Child, although they had heard of it. They knew it was an unusual and potentially dangerous Child. One that could be a vessel of great hope or one of great misery. Judging from the Lord, the Red Child had chosen the latter.

Exchanging glances, Mer and Zel motioned the Holy forward. They approached the Red Child and began to pepper her with questions. When they were through, they rendered their report to the Lord. The Lord held his head low and rendered his judgement. As fearful as he was of the Red Child, he had always held out hope. He had wished deeply that his example would inspire the Red Child to compassion, but he realized that as long as he wore the cloak of the Lord, few would witness his true nature. There was an uproar from the Damned who had followed the Red Child.

Their allegiances were conflicted between the Lord and the Child. This immediately changed the demeanor of the Lord to his former fury. With a sweep of an arm, a whip of searing flame split the Damned into two factions and violently slammed the mutineers to the center with the Red Child. The Lord then said, "judge them all together, for they are all damned." And the judgment was read, and the Lord was charged with executing it.

With the Red Child and her assembled hoard of Damned wrapped in laces of flame, the Lord ascended into the chromosphere of the sun. This is the outer layer of hydrogen and helium gasses. It was rare that any Child of the Sun would ever ascend to this place. It was cold and hostile, the very edge of their world. To stand on the surface was to invite fear of falling into the galaxy. As they broke into the atmosphere, a silent awe swept over all of them. Inside the Star is a life of forever light. Intense luminosity that nothing escapes. Here, at the edge of the Sun was a vast plain of darkness. Even with the sun's brightness all around them, the inky black of the outside universe was breathtaking. Pinpoints of light in the distance were the planets which revolved around them, but they had no knowledge of such things. They only understood the vast void in front of them.

It was here that the Lord proclaimed them all traitors and criminals. They had been judged and found guilty. Their sentence was an eternity in the vast cold of space. The winds began swirling at the feet of the Red Child and the Damned who followed her. It rose in fury until it was a veritable storm. Raising his arms high, the Lord coaxed the flames higher until a solar storm had engulfed all. The Lord lowered his arms and looked at the Red Child, whose face showed no sign of emotion. With a fierce yell and raising of arms, the Lord brought forth a prominence of massive velocity. A massive arch of flame reaching into the heavens. It wrenched the guilty from the surface and thrust them into the atmosphere. Shot as if from a cannon into the very Corona of the Sun and into the blackness of space beyond.

The prominence was searingly hot, even to a Child of the Sun. It took immense power to will it into being then to direct it to excommunicate the Red Child and her own Damned. The skin melted from their bodies and was crusted over, then re-melted again and again as the prominence ejected them. As the Corona subsided and the group entered true space, the Red Child went from the pain of searing heat to the infinite agony of absolute cold. Not the cold of ice or snow, but the absolute freezing of space. In human terms, the core of the sun is 27 million degrees Fahrenheit and the outer Corona is over two million degrees.

The Children of the Sun easily traverse this wide spectrum of temperatures, but space is -454 degrees Fahrenheit. This is not only a lack of heat, but a freezing that is absolute and complete. The abrupt switch from molten to frozen was cruel and confusing. Both burn in different ways. One, familiar, one entirely foreign and excruciating. As her ruby-red skin cooled to a concrete gray, she understood the true meaning of consequence. To this point, the centuries had taught her only moving forward and growth. In this moment she learned a valuable lesson about freedom of both body and mind. In this moment, her body became captive to the cold and her mind to the captivity of coherence. She could hear, see, and feel everything, but was powerless to move, her body frozen solid and weakened by the cold.

Eternity is infinite. Within the field of vision, the Red Child could see only pinpricks of light. Her velocity had declined over the centuries in space, reducing the movement of the lights to a crawl. In space there is no sound. The silence is deafening compared to the constant hissing and explosions of gas on the Sun. Although she never shared her thoughts, memories, or emotions with anyone on the sun, the absence of conversations, instructions and movement were painful to her. Drifting endlessly. No sense of time to measure the passing of centuries. Occasionally her body would complete a rotation and she would see her beautiful red star moving farther and farther away with the years. The day came when she rotated to see the planet Mercury. It was a grey, lifeless orb floating in the inky black. The sun reflected off the glass surface, painting the gray with the red tones of warmth. It sped past her field of vision as it rotated around the Sun, like a speck of dust on a turntable, it appeared on the horizon, passed in front of her eyes, and disappeared, only to return. She would eventually pass the planet by and see only the dark, black side of the planet come and go.

Within centuries she would see the red of Venus appear in the distance. It rotated opposite of Mercury. Its surface rippled with heat waves, smaller and less intense than those of the Sun, but a reminder of her home. She could almost imagine a warmth emanating from the planet, but it was an emotional warmth, not a physical one. The planet moved in line with her own trajectory, and she feared being crushed by the planet as it sped by her. Once again, as she passed the planet by, all that was left was a grim, gray to black in the shadow of the Sun, fading to a dim flickering light. The rotation of her body was interminable. It was barely perceptible except as measured in decades or centuries. The day came when she saw a blue planet in the periphery of her vision. It came in and out of her field of vision as it rotated around the Sun. It was beautiful, comfortable, and hopeful. Such cold colors, but they called to her. She watched the planet over decades as it crossed her field of vision, and she watched it until the last moment where it passed her. Each rotation it came closer, but the centuries had taught her that her fate was cast, and escape was futile. She had given up. Still, each pass came closer until the day she passed in between the planet and its moon. The next pass, the planet would likely crush her.

The earth is spinning at a staggering 1600km. hour, yet to its inhabitants, this is registered as a lazy passing of day and night. With each day the sun warms the surface, and each night, the sun hides, and the glow of the moon provides comfort and illumination.

After the consummation of Mer and Zel, the side of Earth facing the Sun was badly scorched, and the polar caps melted. Earthquakes crushed landmasses together forming massive mountains. Its oceans were evaporated, and the sun-side vegetation burnt to ash, but the planet was resilient. Most of the animals died from the radiation fallout, but a handful of the dark-side plants and creatures had survived. For years, rains poured down on the globe. It washed the ash into the oceans. It drenched the lava beds, breaking down the lava into sands. The clouds formed a solar barrier bringing fierce winters, but each year the sun found more spaces in the clouds to heat the land, gradually creating seasons and warming the Earth. The creatures that existed underground came to the surface. Over tens of thousands of years, they evolved from cell to fin to feather and eventually skin. In their DNA was the essence of their previous life and it came out in the most amazing and abundant ways. Just as snow melts feeding streams, which join into creeks, that merge into rivers that flow into lakes, so did the evolutionary paths of the animal and plant kingdoms of Earth split, join and coalesce into new forms. Every generation brought something new. It was a burst of evolutionary life that brought humanity to the front of the food chain. A human's size is diminutive, but his intelligence and adaptation make him the apex predator. Each generation made him more adaptive and more deadly. Combined with an inherent bloodlust and lack of emotional control, humans completely dominate Earth within a few thousand years.

Today, humanity congregates in cities teeming with filth and crime. They huddle together for protection, not from the elements or animals, but for protection against each other. They lock their doors and create weapons that both protect and annihilate. It is not the strong that survive, it is the ruthless and cunning. It is not benevolence and happiness that is bred, but fear and subservience. The control of the planet is in the hands of a few men and women who use the world as their personal ego game. They send thousands to their death over petty squabbles, they manipulate markets, governments, and lives for the sheer game of it. For profit. Woe to the human who seeks only a quiet life of meager goodwill and cheer. They are targets for those who would strip them of their very soul out of sheer willfulness. You may feel that I am painting far to bleak a picture of the human race but look around. This is our reality. This is what we have allowed for ourselves. We do not deserve better because we do not demand better. As our world crumbles around us, we fight over crumbs of a world that was once beautiful.

†††

It was into this world that the Red Child fell. Drawn into the gravitational field and jerked downward like a ragdoll. She began spinning around the blue planet at incredible speeds, each closer than the last. She took a glancing blow off a satellite, sending it rocketing off into space like the clashing of billiard balls. She felt the heat of friction as she bounced off the atmosphere like a rock skipping on a pond. It felt wonderfully warm compared to the centuries spent in space. She waited for each skip with anticipation of the friction heat it would create. She would eventually pierce the atmosphere and, like a meteorite falling towards Earth, the intense heat of friction revived her frozen body. The friction of the atmosphere against her body heated her until she was glowing red and white hot. For the first time in her life, laughter burst out of her mouth, and she breathed deep of the oxygen rich air, achieving a high that she never wanted to leave. She free-fell with the control of one who was born to ride thermal waves. As the ground rose to meet her, she attempted to pull out of the dive, but grossly misunderstood gravity and the laws of Earths nature. The impact of a soft body on our planet does not create the same destruction as a solid object, but it is devastating, nonetheless. Her limp body slammed into the earth at terminal velocity, driving her form at a steep angle, thirty feet into the earth and ejecting a crater some ninety feet wide. The force of impact shook the earth for miles around, sending tons of desert earth high into the night sky.

Tobias Burlando woke with a start. The blinding light woke him, and the blast sent him scrambling from his sleeping bag as rocks and debris rained down on him. He scurried under his truck to escape the earthen missiles. His old Chevy truck was being battered with rock and boulder, some hitting with force to bottom out the shocks of his truck and pinch him under the weight. In the inky blackness of night, he was absolutely startled. With no reference points, no moonlight, and no understanding of what was occurring, he waited until the last rock fell then slid out from under the truck into a cloud of desert dust. Unable to breathe in the dust cloud, he pulled his grimy, sweat-stained t-shirt over his mouth and nose. In the dark of night, he ran his hands over the body of his truck to locate the door. It was pitted and pounded into oblivion. Whatever had caused the blast had destroyed his old truck. As he reached the door, he was careful to avoid the shards of glass from the windows. He opened the door, and the dome light came on, illuminating the interior of the truck, which was littered with glass and debris. "Unbelievable," he muttered as he searched the floor for his flashlight. He found it and flicked it on, illuminating the area around him. He walked through the dust cloud, assessing the destruction of his truck. The roof and hood were caved in, the windshield and window glass were blown out. The bed of the truck was half full of rocks and debris. The tires looked good, and he assumed the engine would start.

His old Catahoula hound, Jenga, was shaking as she came out from under the truck. This is where she slept when they camped in the desert, and she was not anxious to come out now. Normally, she would chase and charge any animal that was within sight. She was a coyote killing, cow chasing, rabbit eating dog, but this was entirely outside her comfort zone.

Tobias looked around the vehicle and noted that the dirt was predominantly coming from the eastside, so he assumed that the explosion had occurred from the East. He crawled over the driver's seat into the back and found a bandanna. He drenched it in bottled water and tied it around his face. It is common to leave your clothes in the cab of a truck during desert nights to keep creepy-crawlies from nesting in them and he found his jeans and boots without trouble. He poured the glass out of the boots and shook it from his jeans before getting dressed.

Tobias stood still, listening intently for any noise. That was his first mistake. The blast had disturbed the nests of both red-ant and scorpion. The ground was teeming with biting insects that were confused and angry. The first ant-bite had him scrambling to take off boots and scramble out of his jeans to scrape the angry ants from his legs. He jumped into the bed of the truck and by the time he was done he had nine painful welts coming up on his legs and thighs. He shook out his jeans extra well to ensure that no stragglers were hanging on. He dressed again, tucking his pant-legs into his socks. He moved quickly from the bed of the truck to the driver's seat and smiled grimly when the engine turned over. With one good headlight and a small flashlight, he leaned out the window and followed the trail of rocks and boulders across the desert.

The Mojave Desert is nearly 48,000 square miles of sand, cactus, and rock. It stretches from Arizona through Nevada and into California. It is one of the most beautiful and hostile lands on the planet. It is also home to the China Lake Naval Center, where military missiles are tested, and Edwards Air Force Base, home to every American warplane in existence. Tobias was prospecting his favorite opal mining grounds at the epicenter between the two, a mere 30 miles from each. An hour's desert drive either North or South and he would be on military land. He preferred to stay clear of the government, yet he often broke this rule by prospecting in the State Park between the two.

He knew two things right now. One was that a massive explosion had occurred remarkably close to his camp, and two, was that this was certain to bring the military down on top of him. Perhaps it was even a stray missile from a military exercise? No doubt he would be taken in for questioning and could be held for some time. He was not keen on being a prisoner of his own country, but his curiosity was high.

Roll a rock across the sand and what do you see? You see a trail carved into the dirt leading to the rock. Tobias need only follow the deep creases in the dirt carved by the thousands of rocks and boulders that dotted the once flat desert sand. Jenga stood in the back of the truck, feet perched on the sidewalls, head peering around the cab to see through the choking dust.

Tobias could not have missed the steep drop-off into the crater for it had a high berm from the direction he came. His front bumper hit the berm before he saw it and he nearly bit his tongue with the force of the impact. Jenga flew out and over the bed of the truck, landing on her side and rolling in the dust, before coming to her feet and shaking off. Tobias laughed at his initial anger at the new dent in his bumper. It was probably the least dented part of his truck now.

Tobias backed up and drove slowly around the crater, looking into the abyss below and not knowing how deep it was. Eventually, he had circumnavigated around to the impact entry, a long slope carved downward into the bottom. Stopping, he exited his truck amid the swirling cloud of fine desert dust. He dug though the dirt in the back of his truck, unearthing his pack, a pair of leather work gloves and a mattock for digging and scraping.

Tobias edged down the ramp carved in the dirt; down the long, sloping, sandy cone that was freshly hewn from the desert floor. Jenga followed him cautiously. Tobias searched the bottom of the crater for a telltale sign of a meteorite. That would be an amazing find, perhaps even creating a profit for his otherwise meager earnings this year (and many years before that). He would need to work fast to recover it before the military arrived.

The ground in the bottom of the crater was hot. The heat

was making the rubber on the soles of his feet warm and sticky. Jenga was trying to approach the crater center, but the heat on her paws was too intense and so she ran back and forth in the cooler area, clearly agitated and wanting Tobias to get to work. If he were lucky, the meteorite would leave a telltale shiny or charred surface for him to spot. Mattock in hand, he dug in the earth, alternating between the broad blade that quickly removed material and the sharp end that pierced the crust that had formed with the intense heat of the blast. He worked around the outside of the center, not wanting to damage the valuable meteorite with the tool, rather, he allowed the dirt in the center to fall into the trough he was creating, slowly eroding away the center to reveal what he hoped would be a treasure. He anticipated striking a hard-metallic object, so imagine the surprise when the flat blade of the mattock struck something soft and malleable. To one who uses tools regularly, the information that travels up the tool handle to your hands is invaluable in understanding what you are excavating. It determines how hard you strike and where. This soft impact immediately translated to him that he was digging up a body. His senses immediately repulsed before he realized the absurdity of the thought. Perhaps a coyote or other animal was buried in the blast. He knew that his blade had not struck rock or sand.

He gently used the flat blade of the mattock to pull away the dirt from the source of the soft strike. Sure enough, he saw a patch of dusty crimson skin. He knelt and touched it softly and a scalding heat burned through the thin glove, raising a blister on his well-calloused fingers. He cradled his fingers in his hand as he tried to soothe the burn that would ultimately form a blister. With care, he used his hands to pull away the hot sand from around the form, revealing not an animal, but a human leg. Panic gripped him as he began to frantically pull the dirt away from where the head would be. In short time he had revealed a bright pink head; hairless and smooth. He continued to excavate to reveal the soft feminine features of a woman. There was no scratch or bruise on her despite being buried at the center of a blast. Tobias did not stop to ponder this question as he continued to unearth the form. His hands were blistered and burned, skin peeling from inside tattered leather gloves as he dug. A most beautiful woman, completely naked, her pink skin shone like opals where the dust was clear. Tobias was talking to himself all the time he was excavating. "Please be ok, please be ok," he said to himself. He knew he must get her to a hospital, but she was still too hot to the touch. He removed his jeans and wrapped the heavy denim of the legs around her arms and sat her upright. Her head rolled to the side, and he braved another burn by checking for a pulse. There was none. He sat her upright and raced up the earthen ramp to his truck. He opened the rear door and gathered up his sleeping bag, hauling it down to the girl. He unzipped it and, with a fresh pair of leather gloves, worked her into the bag and zipped it tight. He then lifted her, bag and all into his arms and began the uneven ascent up the ramp to the truck. The sand required that he lose two thirds of each stride as his feet pushed through the grains. It was a tough climb, but he lifted her into the rear bench seat of the truck, loaded Jenga in the bed of and started for the driver's seat.

The jets were flying low and fast as they blasted past the crater site. He could not see them until they passed, but they were flying low over the crater, and he knew they were just an advance team for others that would come. He dove into the truck and turned off his lone headlight. Using only the flashlight held close to the door, he slammed the truck into gear and carved a wide arc around the crater, fishtailing across the desert towards the dirt road that would take him to the highway. He should have stayed until the helicopters arrived and turned the girl over to them. Surely, she needed medical care, but something in his heart told him to run, and he did.

†††

The two jets had climbed in altitude following their fly-by, then banked hard to the right and made another pass. In those few minutes, Tobias Burlando, an old prospector looking for treasure in a solitary desert, had slipped out through the dust, and emerged well outside of the pilots focus on the blast site, as they made a second pass over the area.

The pilot's night vision systems were obscured by the dust that efficiently camouflaged Tobias and the truck. On their return pass, they used their thermal imaging systems. In the intensity of heat from the blast site, they never saw the tiny heat signature of a truck that had only been running for a few minutes and had now reached the highway, blending into the other sparse traffic that traveled the Eastern Sierra Nevada Mountains.

They radioed for helicopter and ground support as they circled the blast site until a more appropriate aircraft could land on scene. It would take the helicopters 10 minutes to scramble and another 15 to arrive on site. Two landed at the blast site while the third circled. Tobias knew that in the open desert he was a sitting duck, waiting for the hunters to swoop in and claim him. He had headed for the highway and made a last-minute decision as his wheels hit the pavement. Highway 395 ran north and south from Los Angeles to Reno, following the eastern slope of the Sierras. If he turned left or right, he would be on a major highway, in a truck that would attract attention. Every driver that passed him would remember his truck and his direction of travel. If he continued straight, he would cross over the highway heading West and travel on old dirt roads into the desert lands allocated to off-road enthusiasts called Dove Springs. On the far side of that boundary was the foothills of the Paiute Mountains and a four-wheel drive road that would take him over Bird Springs Pass and into the Kern River Valley. It was a little-used four-wheel drive road, and it would not save him any time, but it would take him far off the beaten path. Hopefully, an area they would not think to search for some time.

He gunned the engine to cross the highway, then passed through a cattle-guard entry that led to Dove Springs. This time of year, there were always hundreds of RV's and pick-up trucks filled with dirt bikes and desert racers. No sooner was he on the road than headlights of motorcycles and trucks began coming towards him. He rapidly turned on his one good headlight so they could see him as well. It seemed the blast had woken the whole camp. When the jets flew over-head the curiosity of the off roaders overwhelmed them and they wanted to get in on the action. They were heading down the hill to gawk at the aftermath. This could be a blessing, if they created a distraction, or a curse, if they told the author-ities of a beat-up truck driving up from the blast area. Tobias continued West, moving to the side of the road as the riders and trucks continued in a stream. This was now an official "event," and no one wanted to miss the party. Tobias only hoped that the crowd would provide an effective smoke-screen for his exit.

The road up to Bird Springs Pass is a bumpy, rutted, narrow passage, once used by pioneers coming over the Sierras to the gold fields. Tobias' old Chevy bounced and jostled its way out of the desert floor, into the chaparral brush and eventually up into the thick forest of the Paiute Mountains where he would be difficult to spot from the sky. As they climbed into the high country, the temperatures dropped. Without windows in his battered truck, Tobias hoped that the cool air would cool down his passenger and be a comfort. Instead, it seemed to agitate and discomfort her. His wheels slipped over rut and rock, and he climbed up into the mountains at a pace dictated by the condition of the road. At 5300' elevation he stopped under a canopy of Jeffrey pine. He shut off the engine and exited the truck. In the distance he could hear the jets circling and see the helicopters searchlights as they played across the desert below. He watched for a few moments and realized that none were searching the road he took. For now, they were safe.

Tobias returned to the truck. He strapped his small headlight over his ballcap, turned it on and opened the rear door. The girl had been tossed about on the dirt road and was now wedged in between the front and rear seats, still nestled in the sleeping bag. Swearing to himself, Tobias moved to lift her into the back seat. She was crimson red and extremely hot to the touch, so Tobias donned his leather work gloves and gently lifted her, sleeping bag and all, into the rear seat of the old truck. What mystified him was that she did not seem to be breathing, yet her skin was supple and hot, like she had a fever. She was clearly agitated, and her body flinched when he touched her.

Illuminated in his headlight she was beautiful. Her skin was radiant and smooth. Her features strangely fluid and unearthly. Again, he noticed that there was not a scrape or bruise on her. Maybe she was a space-alien, he chuckled. At least he was keeping his sense of humor. Tobias had removed most of the dirt and rock from the back of his truck, unearthing his other clothing and water. He tried to give her a sip of water and she coughed and sputtered it out, her face contorting into a mask of pain, then she went still again.

Tobias swabbed out the dirt from Jenga's travel bowls and filled them with kibble and water, then he then turned to his own wounds. Removing his pants, he fished around in his center console for a tin of OHO Mojave balm, which he applied liberally to the ant welts, and his hands, which were severely burned and sensitive. He opened a second tin of OHO mesquite balm to prevent his burned and blistered hands from cracking. Old Native American remedies always worked the best.

He found an old pair of soft cotton work gloves in the door storage of the truck and slid them over his hands to protect them. He checked on the red girl one more time before he turned off his light and sat with her as he waited for morning. Helicopters droned in the distance through the night. Tobias worried about her fever. He worried about her color, and the way she had been at the center of a massive blast and survived. What was she doing out in the desert? He would have seen if there was another vehicle or prospector out there. Something was happening with this girl, and he felt the need to protect her. He knew they would have discovered his truck tires and the hasty excavation. He had hoped that the discovery of the mattock tool he had left at the site would satisfy them that he was just a treasure hunter that was spooked off by the jets. He tried to remember if there was anything else that could link him to the site. He would need to be more careful from here on out. Tobias ate a crusty granola bar he had found under the seat of the truck, and he drank one last bottle of water, leaving one for the girl when she woke and one for Jenga, who was off chasing mice and coyotes in the night.

In the pale dawn of morning, Tobias unzipped the sleeping bag to check on the girl. She was visibly paler, but still a very bright pink color. She was hot to the touch, and it pained his burned hands to handle her, but he dressed the girl in his spare clothing, surprised to find she was nearly as tall as he was. Tobias was no stranger to a woman's body, but he found himself embarrassed by touching this beautiful young girl. He zipped up her bag and she visibly relaxed, obviously enjoying the heat. Tobias dabbed her lips with water, and she again recoiled as if burned by the moisture. He resolved to not give her more until she was awake. They sat all day under the canopy of trees and Tobias would walk to the overlook, gazing down into the desert floor whenever he perceived an increase in the sounds. The crater site was now only visible by the mass of vehicles and helicopters flying in and out. Tobias could see the red and blue lights of police blocking the highway to the north and south, turning traffic around and sending them back the direction they had come from. Large crowds had formed behind the roadblocks as television crews, rubber-neckers and UFO enthusiasts fought to get a glimpse of whatever fell from the sky that would attract this much government attention.

By nightfall, the last of the helicopters had droned into the distance and the trucks stopped patrolling. Traffic had resumed, but the crater site was obviously still under a quarantine. Tobias could see a large tent structure had been erected over the entire site and several vehicles were still gathered around. Tobias determined that he should leave his perch and head down the other side into the mountain valley in the early hours of darkness.

The lone headlight provided little illumination as they drove the winding dirt road down the backside of the mountains and into the Kelso Valley, eventually reaching the paved highway in Weldon. Tobias turned left and set off down the road. He imagined every car he passed gawking at his battered and beaten truck, although it was dark enough, they would likely see little. He turned right at Sierra Way and soon Lake Isabella was on his left as he drove the shorefront back-road to home. By the time he rolled into Kernville dawn was nearing. He pulled his wrecked Chevy right to the front doors and carefully carried the girl inside with Jenga following. He walked her to the downstairs bedroom and lay her, sleeping bag and all, on the top of the covers. He returned to the Chevy and moved it to the back of the house and under the carport where it could not be seen from the road or the sky. He filled up the water and food bowls for Jenga before grabbing a washcloth which he wetted with cool water to clean off the dust and grime that covered the girl. Upon contact with her skin, she let out a cry of pain and he immediately stopped. He returned to the bathroom and filled a small pitcher with steaming hot water, then went upstairs to his room to fetch his winter ski gloves. His hands were angry and red from the burns, and he could not handle a hot washcloth without protection on his hands. With winter gloves, he dipped the cloth into the hot water, wrung it out and wiped it across her skin. She sighed as the heat soaked into her and rose into steam. Each wipe of the cloth revealed a skin as pink as a pig's nose and as shiny as a polished opal. She was stunning. He began removing the clothing and cleaned her entire body, returning often to keep the water scalding hot. When he was done, he fetched his warmest flannel, dressed her, and pulled the down comforter tight to her chin. Exhausted, he went upstairs and took a hot shower, shaved his stubbly chin, and crawled into bed. He was asleep before his head hit the pillow.

Tobias Burlando opened his eyes to a checked blue form standing over him. Startled, he scrambled away and out of his bed on the opposite side, landing on his feet and waiting for his head and eyes to clear. Before him stood the girl dressed in checkered blue flannel. Her face a stony blank slate, her pink head radiating light and her huge eyes a kaleidoscope of colors. Eyes closed and helpless she looked the part of the injured little girl. Standing before him now was a powerful force who was clearly not human at all. The intensity of her gaze held neither anger nor cowardice. It was impassive and cold. Tobias held his hands out in front to show that he had no weapon, that he was not a harm. "M-my name is Tobias," he said. "I found you in the desert." The girl stood looking, not betraying an emotion. "I'm glad you are awake..and, er, well," he stammered. "Are you ok?"

The girl looked at Tobias as if determining his value, then turned and left the room. Tobias scrambled into the hall and to the kitchen after her. She looked around at the room, then out the large picture window to the world outside. She walked to the window and stared outside for a long time. She reached out to touch the glass and found it solid. She pushed it gently and it flexed slightly, "NO!" shouted Tobias, a little too loudly. She turned to him with the same steady gaze. "It will break," he said more gently. "Would you like to go outside?" Tobias crossed the room to the door and turned the knob, showing her how it was done. He walked through the doorway into the brisk morning air. The girl came to the door and felt the chill in the air. Shivering and shoeless, she stepped outside and felt the cool breeze on her skin. She walked to the deck railing and stared out at the surrounding homes, the mountains and the blue sky dotted with clouds. It was so different from the world of gas and flame she knew but was still a welcome relief from the freezing cold of space. She did not know how long she had floated out there, vaguely aware that her Damned were slowly drifting in different directions and scattering to the universe. Somehow, she had arrived here in a place that was neither hot nor freezing. The coverings she wore provided small comfort and she burrowed deeper into the flannel. Tobias went to the door and beckoned for her to come inside. He led her back downstairs and carefully built a fire in the rock fireplace for her. He noticed an immediate change in her when the match caught, and he touched it to the wax and pulp firestarter. She was mesmerized as the flames licked up the sides of the logs and began to char them. She reached out her hands and drew the fire towards her, willing it into her body. The heat was brief but robbed the wood of the heat it needed to continue burning and the fire went out. Tobias had been watching with fascination as she literally sucked a flame into her body, illuminating the pink skin from the inside, albeit briefly. He saw the pleasure on her face as she drew it in and the frustration when she extinguished the fire.

Tobias quickly rebuilt the fire and stacked it high in the fireplace. He urged the girl to wait a moment for the fire to completely catch. She desperately looked at the flames wanting to claim them but understanding that she must wait.

When the logs were completely consumed by fire, Tobias motioned for the girl to gently take in the fire. Not too much too fast. Once again, she reached out her hands, immersing them directly into the flame, and slowly sucked the flame into her, modulating the draw so that she was nourished without completely extinguishing the flames. Tobias fed log after log into the fire as her skin glowed from pink to bright red. The room became blistering hot, and Tobias was sweating profusely. He went through the house opening doors and windows, but in the still air outside, it made little difference. The wood floor where the girl stood began to char and it was then that Tobias spoke, "you must stop," he said. She looked at him and he gave a signal to halt. She lowered her hands and the flamed died down. Tobias fetched a pitcher of water and doused the flames enough to keep it going without burning his house down and spilled some at the girl's feet, which hissed and popped as it vaporized.

For the girl, this was the first time in memory that she could remember feeling warm again. So many years drifting in the freezing vastness of space, a brief respite as she thawed on entering Earth's atmosphere before settling into the uncomfortable chill that permeated her body. She felt reenergized by the brief warmth of the fire, and she knew she needed more if she was going to survive here. For now, she was sated.

She stayed close to the fire and Tobias piled his old wool blankets over her to keep her heat in, and to make the room cool enough for him to stay inside. He pondered this new reality. Since she opened her eyes, he had felt certain that she was not in the crater because of her proximity to the blast, but that the blast was caused by her. She had fallen from the Stars, and he had found her. Was she an Angel falling from heaven, or a Demon falling to Earth? He still did not know if he was protecting her, or harming himself, but he trusted his intuition and was determined to observe and assist where he could. What the last hour had taught him was that she needed intense heat. Heat far more powerful than he, or his house, could stand. She would need a source of heat to warm her and an insulation to keep her warm. For now, she seemed comfortable.

The girl read the signs and was coming to the same conclusions. When she first opened her eyes and saw this creature, she was puzzled to see how alike and how different he was from herself. The first thing she saw was the strands of fiber covering his head, and his eyes were small with concentric circles within them. His skin was brown. He was much smaller than the Children of the Sun and only slightly taller than herself. He showed his teeth a lot and it was a pleasant and reassuring expression. She did not fear him for she had never known fear, but there was something else here. She trusted him. She could feel his own apprehension during their flight from the desert. She could feel that there was danger for both she and him. She trusted that he was guiding her to safety, for she had no one or nothing else to connect to. She would watch and learn.

Mostly she was puzzled by the sounds that came from his mouth. It was entirely unlike the sounds her own people made, and she could make no sense of it. Surely it was a form of communication. On the sun, communication is found in gesture, sound, touch and even through connecting minds. The heat itself is a communication. Communication had been with her always and she had never thought about how it happened; it just did. In this place, the air did not communicate. Aside from various facial expressions, the body did not communicate either.

She reached out with a hand to touch his arm and he recoiled quickly as the heat of her body once again burned him. She saw the skin turn black and split open, she smelled the burning of flesh and his alarmed and pained expression. She realized that her touching him had hurt him. Looking at his blistered and charred hands, she realized that in saving her, he had been badly hurt himself. Still, in that brief touch she had connected mentally to something. She willed all the precious heat away from her hand, which diminished in color from scarlet red to pink as she held it for him to see. When it was nearly white, she was obviously as uncomfortable as he had been when she had burned him. She slowly reached out to touch him, her gaze reassuring. When her warm hand touched his arm, he felt the warmth seep clear thought his body and flood his brain. He knew that she was physically connecting to him, and he stayed very still. The girl sent tendrils of thought throughout his body, looking for a means of communication. She found it inside his skull, in a gelatinous mass. The amount of information it contained was staggering and the girl began to sift through the life of Tobias. His experiences, his learning, his family and his knowledge of his own body and environment. She also saw how he saw her, as a woman, a female. She was unaware of gender differences and the special connections between the male and female of this world. On the Sun, these feelings and experiences were reserved only for Mer and Zel. On this planet it seemed procreation was practiced at will. When she had what she needed, she sent a single thought back to Tobias. "I understand." She then let his hand free. Even with her cooling of her hand, it still left an angry pink "sunburn" in the shape of her grip on his wrist.

Tobias was perplexed. He knew that the girl had permeated his brain, that she had read it like a book. He felt violated and used, but he also understood that she needed information and that was the easiest way to get it. Still, he had many questions himself. Foremost was, what was he going to do with her?

The answer came into his head like a voice. "Greetings Tobias. I am known as the Red Child, one of the Damned from the center of this Solar System. Do not fear me for I do not bring you harm."

Tobias looked to the left and right, looking for who had spoken. He was sitting right in front of the Girl, and she had not said a thing. "I have determined that I can connect to your mind directly," said the disembodied voice. "I lack the means to communicate verbally with you. I hope you will accept this form of communication."

Tobias looked directly into her kaleidoscope eyes and thought "what are you?" And Tobias' mind was filled with visions and memories of the Red Childs life in the Sun. She withheld the executions, and her own trial, but showed

Tobias her ejection from the Sun in a powerful solar flare. He felt the icy bitterness of space and her fall into Earth's atmosphere. The next thought was of her opening her eyes and seeing him. And he understood.

For several hours they discussed the situation. Tobias determined to call her Red, which she accepted. He shared a vision of the military and of conspiracy theories of government scientists cutting up Aliens to study them. He shared his intuition that he was meant to protect her. This amused the Red Child, this soft and fragile being protecting her. Still, she found it reassuring that he would guide and protect her as she learned this world.

They discussed the best way to keep her heated and how to insulate her properly. Understanding that she was born to temperatures that could melt the entire planet was daunting but understanding that she could recharge and sustain comfort was useful. Tobias ventured that a blast furnace used for glass blowing would likely be helpful. He also felt that the suits scientists used to get close to volcanoes would likely work both ways. It insulated them from the outside heat, but could it also keep that intense heat in, insulating Red from the outside cold? For how long? The next question was simply one of economics, for Tobias Burlando was a poor man. He had an old destroyed Chevy truck and this home that was left to him by his parents. No savings and meager food in the house. He made his living searching for gems in the desert. What he made in stones barely paid his way through life. It was honest work; he was able to dodge the majority of taxes through cash sales and he was his own boss. For all of this, he was still flat broke.

Now that they knew a telepathic connection was possible, Tobias allowed her free roam of his mind. Closing his eyes and feeling the ticklish movement of synapses firing and bursts of memory in rapid fire. He had no hidden secrets or information that could harm anyone. Frankly, most of his information came from the television or internet, which was entirely unreliable. Through this wandering of his mind, Red understood commerce, and finance in its most rudimentary fashion. She knew that they would need money to purchase the things she needed to survive on this planet. The best way to achieve that was to give Tobias something he needed to trade so they would not be too obvious. Second, she needed a better source of information on how heat was created on this planet. She needed true intense heat, not the subtle warmth of a fire. She also needed to collect this heat without putting Tobias, his animal, or his home at risk. A short run through his mind and she told him to fetch the electric burners from the garage. He had completely forgotten about them, but there they were, hidden in his mind. He left the room to fetch the burners and Red stacked the fire once again with wood and began to syphon off the heat at a rate that allowed the fire to continue.

Tobias returned with two small electric burners. The cords were frayed and the burners dusty, but they should work. Once plugged in, the old coils began to glow red, putting off meager heat for a room this size, but perfect for her hands, fingers splayed, to rest atop the burner to absorb this heat. Red rested her hands directly on the burners, channeling the heat through her body. This was not ideal, but it would do in a pinch. It allowed her to remain bundled and wrapped to preserve the precious heat her body accumulated.

Tobias had a thought. If electrical energy could make this coil hot, could more energy make larger things hotter? Flipping on his computer, Tobias searched electrical heat and discovered that it was incredibly efficient and hot, but the shock to achieve this heat was generally far more hazardous than the heat itself. He devised a simple test. Once Red had sufficiently warmed herself, he disconnected one of the old burners and with a screwdriver, removed the backing plate. He removed the screws connecting the frayed power cord to the plate and then had Red hold one bare wire in each hand. He warned her of the potential shock, and she sent back a reassurance. Tobias plugged the socket into the wall and the lights in the room dimmed briefly then went out, the fuse clearly blown, but the test was a success. Red excitedly relayed that she had received a wonderful and warm reception from the wires. Not enough, but better than the fire. They were going to need a high voltage test.

Tobias walked up the street to Slims house. Slim was anything but. His six-foot four frame supported nearly 400lbs of weight. He was the most cheerful man Tobias had ever met, and he had several rusting cars in his yard that he claimed to be working on. Tobias walked up to the door, knocked, and heard the lumbering footsteps on the wooden floor. The door swung wide, and Slim was looking down at Tobias with a huge grin. "Tobias, welcome buddy, what brings you 'round," said Slim.

"I'm afraid I don't have time for a visit Slim. My truck was damaged in the desert, and I need to borrow a car from you if you can spare one for a few days."

Slim surveyed the assortment of cars in the yard and told Tobias to hold for a second. Slim went inside and soon returned with a set of keys for the old Oldsmobile. "She ain't much to look at but she will get you down the road," said Slim.

"Thanks, I will bring her back with a full tank," said Tobias.

"Don't you worry about the gas, use it up. I'm going to drop the tank when she gets back anyway. Better that she's empty."

"Can I bother you for another favor?" asked Tobias. "Can you watch Jenga for a few days?"

"Sure thing brother," said the large man.

"I appreciate it," said Tobias as he retreated down the steps, waving as he went.

The Oldsmobile was an ancient rusting heap. It was all of 30 years old and those were hard years. Slim could get her running, but she would never be restored, and that's a fact. Tobias turned the key and the engine jumped to life. A big, hulking v8 from a bygone era. Tobias threw it in reverse and backed out of the weeds and onto the driveway, reversing all the way to the road, then dropping her into drive and setting off down the hill to home. These old vehicles drove like land yachts and the old springs and shocks set her at a lazy wobble down the street. Super comfortable, but a long distance could make you seasick.

Tobias checked the fuel level then headed down the highway to the Army/ Navy store twelve miles away. He went through the piles of winter clothes that were on sale as spring approached. He was collecting anything that looked like it could insulate Red without combusting. He found a pair of sturdy boots for her feet and double thick cotton socks to keep the heat in. He managed to dig a thermal suit used by iron workers that was sized two times too small, but perhaps he could modify it for her. He found some Nomex fireproof undergarments used by forest service firefighters. He went to the glove aisle but found none that would resist melting to her skin. He did buy several pair that would work well for him to handle her, and a few pairs of rubberized power-utility worker gloves for handling high tension electrical lines. If they were going to monkey around with hot electrical wires, he was going to need to be extra careful. He found a large pair of women's sunglasses that were extra dark and would completely cover Reds eyes.

He paid for his purchases and walked down the walkway to the end of the strip-mall. There was a small hair boutique there that catered to the old blue-haired ladies of the town. He peered into the window and saw four sets of eyes staring back. He also glanced what he needed on the back wall. Walking inside he said, "good morning Miss Betty, Miss Lilah, ladies." He was all smiles and charm, trying to keep the conversation moving forward rather than let these women hen-peck him. Miss Lilah started to ask her first personal and inappropriate question when Tobias turned his attention to the far wall and inquired of Miss Betty if the wigs were for sale. "Why yes, sugar," Miss Betty said in her most sultry Southern drawl. "What in the world you want with a wig?" Before Miss Lilah could get a word in, Tobias said something about a costume party in the city and needing a wig. Miss Betty said that it was gen-u-wine human hair and that it was expensive, but Tobias persisted, and she named the price of $50.00. To the amazement of the women present, Tobias pulled out his credit card and asked Miss Betty to wrap it up for him. "Tobias Burlando!" said Miss Lilah, "what in the world kind of party is it that you would spend $50 on a wig for?" Tobias just nodded politely as if he did not understand the question and he waited while Miss Betty processed the card. The women were now a gallery of gossip and questions. Tobias Burlando was known for being frugal, and possibly even poor. He watched his nickels and dimes. Now he was here spending good money on a woman's wig. What in the world could Tobias Burlando be up to? Who was this woman that he was seeing? Tobias thanked Miss Betty and avoided the questions that were rapid fired at him as he made his way to the door. Whew. That was the singularly most uncomfortable thing about a small town. He knew that this story would be retold for years, every time he walked through the village.

With his parcels and purchases, Tobias drove back to the house and lay everything out on Red's bed. "I hope these fit you," said Tobias. "We will need to go outside, and you draw too much attention. We need to make you as human as possible," which sounded reasonable to Red. She stripped off her wool blankets, which made Tobias uneasy and awkward, then put on the Nomex underclothes, then the silver thermal suit. Yes, it was too small, but that just meant that it clung to her in all the right places. It fit like a silver glove. She could have modeled it on the runway. Blushing, he showed her how to dress in human clothes over the suit and put on the boots. They were far too big, but two pairs of socks took up the extra space. The bulky clothing did nothing for her shape but make her look lumpy and unattractive. This would keep her from drawing attention. She walked about clumsily in the boots, preferring to be barefoot. When she was fully dressed, he helped her with the wig, completing her transformation. A pair of dark sunglasses and she could walk through town after dark without a second glance from near anyone, except her face and hands were very pink. Tobias hoped that it could pass for an extreme sunburn.

Tobias locked up the house behind him and Red got into the Oldsmobile while Tobias walked Jenga across the road to Slim's place. He kneeled and buried his face in her fur and ruffled her ears before putting her into the fenced area behind Slim's house. He returned to the Old's and they drove out of town heading West to a remote stretch of road where they could test theory number two. Even at mid-day, this was a lonely stretch of highway. Very little traffic would come this way. Tobias slowed and pulled onto a little used dirt road. He headed directly for the chain-link fence and parked the car to block anyone's view from the main road. Leaving Red in the car, he opened the trunk and removed a large pair of bolt-cutters. He clipped the padlock and let himself inside the fence. The equipment was marked with "high-voltage" signs. It was an electrical transformer that managed electricity for communities around the lake. Tobias reached out with his mind, virtually shouting for Red to join him. She calmly responded that there was no need to shout, she would hear him anywhere he was. He made a mental note to be careful what he thought around her.

Red looked toward the road to insure no cars were coming. She exited the Oldsmobile and walked around the car and into the fenced area. Neither of them knew the first thing about how these things worked, they just knew that a lot of voltage was passing through this very spot. Red tentatively felt around, touching various wires and posts. Nothing. Tobias returned to the car and brought his small tool case. He began opening panels and boxes, unscrewing the fasteners that held them shut until he found one with large copper wires crisscrossing inside. "Looks like this is it," he said. Red looked inside and could feel the energy surging through the wires. It was not hot, just energized. She reached a hand inside, brushing a wire. Nothing. She put another hand inside, taking ahold of a wire opposite and sparks flew. Her skin instantly turned bright red, and her clothing and wig burst into flames. The protective suit was working fine, but the areas where her outer clothing touched her bare skin was alight with fire, and Tobias threw handfuls of dirt on them in an attempt to douse them. Within three minutes she released her hold. She was whole again, fiery red with ripples of heat coming off her. Her outer clothing complete gone and she stood with eyes closed in her shimmering silver suit. Heat came off her in waves. For the first time since she left the Sun, she felt whole. She opened her eyes and waved her arm across the landscape, sending a jet of heat energy through the fence, melting it and leaving a scorch mark across the entire mountainside beyond. She focused the heat on a large granite boulder, and it melted like butter in a microwave oven. Tobias was too stunned to speak. Using this energy had barely dimmed her color, but it was obvious that she would need to recharge if she did this often. The interesting thing was the heat signature directly from her. She had managed to minimize the heat coming directly off her body and store it. Both Tobias and Red recognized this together and Red spent several minutes mentally working out the physics of keeping this heat energy inside of her body. While she managed this, Tobias set about working on a solution to getting her home. In this heightened heat state, there was no way she could get into the car. Long minutes passed. Two cars sped along the road without noticing the car parked down the lane, or the two persons behind it. The lava that had been the granite boulder radiated heat but was no longer molten red. It had melted into a black glassy mass.

It was Red that solved the problem of riding in the car. She simply lifted off the ground and floated. This was her natural state. Without the thermal waves of the Sun her propulsion was slow, but she could manage flight at a reasonable pace. Once again, Tobias was speechless. She did not need to ride in the car, but a flying red woman in a silver suit would certainly be a sight for the locals. Flight was not a solution. She must be able to modulate her heat. With some effort, she was able to direct her heat internally with minimal heat signature herself.

Tobias retrieved the alcohol wipes from the car and wiped down everything they had touched. He replaced the latch and hooked the lock shackle through the gate. It would fool someone from a distance, but up close anyone could see the lock was cut. Up close, anyone would see the fence had been melted and a large boulder turned to putty. No, this was not the perfect crime. It was certain to attract attention.

Master Sergeant Tom McGill was the first to receive the notification of a small meteorite impact 45-miles North of Edwards Air Force Base. As the Chief of Security Police for the country's foremost research and test base, he had seen it all. This was a routine exercise. He requested two F-16 fighters be scrambled as they would be the quickest on the scene. He notified his deputy to contact local law enforcement with instructions to secure the area, but not to enter. Next, he put a squadron of Blackhawk helicopters on alert. Last would be his boots on the ground. This was likely just a meteorite, but its proximity to both Edward and China Lake sites made is worthy of every caution.

Within ten minutes, he was walking across the parking lot towards his office and heard first, then saw the afterburners of the two F-16 Falcon fighters as they lifted off runway 2. His staff was beginning to arrive to the office and the coffee pot was turned on. Their workday had begun.

With a top speed of 1,500 miles per hour, the F-16 fighter jets were barely working when they passed over the crater, their night vision powerless to see inside the heavy dust cloud. They shot past the crater like a bullet, proceeding down the desert valley for some miles before pulling up, banking hard to the East, then dropping back low to the ground for a second pass. Fortunately, the pilots were focused on the crater, not expecting that anyone or anything would be in the area, allowing Tobias and Red to slip through. The jets reported the heavy dust cloud created by the impact and requested additional ground support. McGill's car was running and ready outside when he and his team emerged from the office. They drove down the Edwards flight line to the Helo base, where they passed through two separate security checks before pulling up to the Black Hawk hanger.

In the air, the helicopter banked right and left Edwards airspace, heading to a blank spot on the map just 45 miles away. MSgt. McGill was receiving data and updates on the explosion, including images of the meteorite falling in a streak of light and satellite images of a massive dust cloud over the area.

The helicopters hovered briefly to disperse the dust from the crater, then landed a short distance away to keep the dust low over impact. MSgt. McGill and his team exited wearing their radioactive protection suits and immediately began canvassing the area. It was apparent that someone had been on site, that they had excavated and removed something small from the crater. The telltale mattock, and the deeper footstep trail of carrying extra weight back to the truck was enough to order a wide-area search. As the scientists pored over Geiger counters and forensic investigators took castings and pictures of the tire, foot and drag tracks, MSgt. McGill was reporting to Command that they had a breach, and an object was removed. Within minutes, MSgt. McGill was ordered to stand down, remove all staff from the site and secure it for the Pentagon. Trucks began arriving on scene, roadblocks were set up and soon the entire site was a hive of activity.

It was a four-star general that relieved MSgt. McGill of command, ordering him and his team to wait some 2 miles away at a checkpoint until further instruction. As the initial crew was preparing to leave, armed teams of Navy Seals and Army Delta were landing via attack helicopter. Large Sikorsky sky-cranes were delivering heavy equipment and teams were setting up an environmental habitat and satellite uplinks. Within two hours there were many scientists and pentagon officials working feverishly. It was only a rumor, but McGill heard that it was not a meteorite, but a human body that had cut through the atmosphere in a ball of flame and slammed into the earth, creating this crater. The rumor said that the body was still intact when it was removed.

Tobias and Red were sitting in his living room. His one piece of good furniture, a leather couch he had bought at a yard sale, had a large hole burned in the seat. His wooden chair was charred. Tobias had bucket, pots and pans of water set up strategically through the house should he need to douse a flame.

Through Tobias' memories, Red knew that the abilities that made her normal on the Sun were entirely unusual here. She also knew that they could draw attention. They must both practice absolute caution if they were to survive. Somehow, they must find a place where she could buy the time necessary to figure this out. To buy time, they would need money. One thing was for certain, they could not stay here. They were too close to the crater site and the Government had the resources to track them down.

Tobias went into his room and fetched a large, heavy box. He then drove into town, filled the car with gasoline, and checked the fluids. Everything looked in order. He stopped by Dewey's rock and gem shop and hefted the box through the front door. Dewy barely looked up from his computer. "Yea 'Tobe," he said. "Wassup?"

"You know those gem specimens you've been after me to sell you for years? Here they are." Dewey's eyes raised up over his glasses, his head never moving.

"Whadya talking bout he said? Your good specimens?" Tobias lay the box on the counter and removed the lid. Inside was a rat's nest of newspaper. Both men knew what they were wrapping. "Whassthisbout?" asked Dewey. "These ur yur fav'rites."

Tobias sighed heavily and said, "they are all here. You offered $3500 for all of them two years ago. If the offer stands, I'll take cash."

Dewey bolted off his chair and moved like a man possessed to his office. With trembling hands, he opened his safe and counted out $3500. These rocks were worth many times the asking price and both men knew it. Still, Dewey had learned years ago never to question good fortune. He returned and slapped the money on the counter. Tobias, picked it up, looked Dewey straight in the eye and offered his thanks. Tobias turned and walked out the door leaving Dewey completely confused.

Tobias returned home and packed a bag with his necessary clothing and made a bag for Red as well. He took them out to the car and returned for whatever food he had in the cupboard. Filling a box, he hefted the heavy canned goods and walked them downstairs and to the car. Lastly, Tobias gathered all his blankets and his big sleeping bag, filling the back seat. He walked to Slims house for a visit with Jenga and she ran circles around him in excitement. He gave Slim a bag of kibble and her leash, then gave her a gentle pat and her head. He left his best friend in good hands. Finished with packing, he looked around the street to make certain it was clear, and he quickly called Red to come. She was out the door and in the passenger seat in no time, and they were on the road.

As they exited town, the first military vehicles were rolling in. With an abundance of caution, Tobias turned swiftly across town, crossing the bridge over the Kern River then left, heading North, deeper into the Sierra Nevada mountains. Twenty minutes later he turned right and began the steep climb into the high country of the Kern Plateau. They would be able to drop down the other side into the desert, intersecting the highway thirty miles North of the original impact site then head North. The climb up the hill was steep, ascending 7000 feet in just twenty miles to the summit. By the time they reached the top, the engine was overheating. Here on the summit, they were sitting ducks. Tobias dropped down the other side, coasting until they hit a dirt road on their left. They continued coasting until they found a small meadow with a running stream. Tobias opened a can of peaches and sat eating them while the wind rustled the pines. Birds darted over the meadow, picking off mosquitos and insects. After the high-intensity action of the last 24 hours, it was a welcome break.

Red loved the forest. It was the opposite of her world. Instead of gas and winds, it was a forest of solid trees and a gentle breeze. She lay on the warm hood of the big car, looking up into the canopy and listening to the forest sounds.

Tobias walked to the stream and filled the now empty can with water and returned to the car. Red slid off the hood and Tobias opened the massive lid, exposing the engine, with its myriad hoses and tubes. Steam was still trickling out of the pressure relief lid for the radiator, and it was too hot for Tobias to open, but Red reached over and twisted, enjoying the scalding hot steam that emitted from the radiator. Tobias allowed the engine to cool before starting the engine and slowly adding cool water to the radiator. It took many trips to the stream to refill, but when done, the engine was ready to go again.

They backtracked to the main road and descended into the plateau below them. They began to climb again fifteen miles later and the engine temperature began to creep back up into the red. They could make a left up into the Blackrock Forest Service fire station and lay low, or coast miles downhill into Kennedy Meadows, where they would find more water in the South Fork of the Kern River. The old car was their only transport out of the area, so Tobias turned up the spur road and into the fire station at Blackrock.

With fire season behind them and Fall coming on, the station had been closed for the season. At the entrance was a visitor's center, and behind, a barracks and dining hall. To the left was an equipment repair barn and, in the distance, Tobias could see a small shed that he knew would house the big diesel generator. He drove to the back of the property, pulling in front of the repair barn. He fetched his bolt cutters and within minutes, the Oldsmobile was parked inside and out of view.

†††

MSgt. McGill was debriefed on base and quarantined to quarters before being summoned to the Commanders office. There, he was briefed by a Pentagon official as to what the official report would show regarding the meteorite crater. As the man most knowledgeable of the desert surrounding the crater, and of security policy, he was being deployed to an independent agency as an advisor. Boyd Troutman was his direct report. Boyd had, in fact, been sitting quietly at the far end of the conference table as McGill was debriefed. He watched the man as the situation was described to him. He could see McGill mentally following the terrain and area as the story unfolded. Every detail was being placed into context and filed away for use. Troutman had already read the file on McGill and hand-picked him for this assignment. He was "by the book," but it was a book that flexed with common sense. He would use every trick and wile available to him to achieve his goal within the confines of his orders. McGill was a thinking and acting man. Troutman would need both.

Troutman stood and approached McGill with casual confidence. He extended his hand and introduced himself as "Troutman" and referred to McGill as "Master Sergeant." Troutman produced a manila file marked "Classified Alpha." Before opening it, Troutman advised McGill that his security was being upgraded to Alpha level. McGill had always thought his *Top Secret* was the highest you could attain. He had a Top Secret/RD/Q level access and was SSBI on several files. As far as he was concerned, few people in the country could match the level of access he held as the top security chief of the Military's most important test and research base. "Alpha?" he asked.

Troutman just grinned and said, "a whole new world is about to open up Staff Sergeant. Try to keep up."

The file was opened, and Troutman shuffled through it. He removed a high-resolution photo of the meteor as it plunged to the earth. The light was blinding in its intensity, obscuring the image. Troutman produced two more images which were closeups and enhancements of the first. McGill could clearly make out both a hand and a leg in the images. McGill did not feign surprise, which told Troutman that he already suspected this. "Who told you?" asked Troutman.

"It was just a rumor in camp," said McGill, "I don't put stock in rumor, but I do listen."

Troutman removed another image from the file. This was not from either the jets or the helicopters. McGill recognized it as an enhanced satellite image. The series followed the meteorite from its first observation as an object in the atmosphere and in rapid succession falling until impact and implosion. The next series of images switched to what seemed a thermal vision image, but McGill had seen the dust cloud and assumed it was taken after the dust settled. Troutman offered in response, "this is our latest technology in thermal imaging. It essentially removes the dust from the image." He offered nothing more and McGill took him at his word. In the succession of images, he watched a man and a dog explore the crater, the man excavating and then dragging an object in a tarp or bag to his truck, loading the truck and speeding off, all while the dust was obscuring the site. A second set of images showed what seemed to be the same battered truck on a dirt road heading west over a mountain pass. The images stopped when the truck reached the tree line.

"The object that created the crater was removed by this subject before your team arrived on site. Due to the extreme dust buildup on the front and back of the truck we are unable to retrieve a plate number. We are looking for a similar vehicle make and model now from Nevada to the Pacific Ocean. They could be anywhere."

McGill looked at Troutman and said, "no, they couldn't. A truck in that shape would catch the eye of a state trooper. It looks like it was battered heavily by falling rocks and debris from above. Likely, very close to the blast site. Maybe an off-roader or prospector. He high tailed and hid, so he knows that what he has can get him in trouble, and he is willing to accept the risk. He went home. He is sitting in his house right now wondering what he should do."

Troutman simply said," go find him Master Sergeant."

McGill stopped at his office to issue an all-points bulletin on a white Chevrolet crew-cab truck made sometime between 1989 and 2004. There must be a million of them. He listed the condition as severely damaged, reducing the number to tens of thousands. He left his Deputy in charge and went home to pack a bag. Home was a dingy brown enlisted house on base. McGill had married twice, divorced twice, and determined that women were more trouble than they were worth. If he needed a woman, he would do what any self-respecting military man would do, he would pay for her by the hour. Within ten minutes his bug-out bag was packed, and he went outside to the waiting suburban. It was black with no government markings, just like McGill thought a super-secret agency would use. This was going to be an interesting assignment.

The driver of the suburban drove to the flight-line and showed his credentials to the gate soldier. The soldier swiped the bar code and returned to the car, saluting smartly. The suburban proceeded onto the flight line, turning right when all the normal flight line traffic would turn left. He drove to a small hanger far from the main cluster of buildings and honked. Double doors slid open, and he pulled into the well-lit interior. Inside were two corporate jets and a Super Puma helicopter. As he exited the vehicle a young woman with bronze skin, flashing blue eyes and a crisp business suit stepped lively to McGill and saluted. "Sir, I am requested to escort you to your team."

She was so perfectly arranged that it was almost humorous, but something told McGill that this was not a green recruit, but a highly trained and groomed operative. "Thank you, miss."

"Lieutenant Comden," she said, still saluting.

"At ease Lieutenant. Since when does a Lieutenant salute a Master Sergeant?" he asked.

"Since you are in charge of this mission, sir," she replied.

McGill shook his head. This was entirely foreign protocol to him. The military had an extremely specific protocol for rank, and he had never heard of an officer saluting an enlisted man, even in a situation like this.

He was ushered across the hanger to a dingy office door. As he stepped inside, he saw that this club definitely had perks. The carpet was thick and plush, the interior expensive and lush. Everything shined like a new penny. They walked noiselessly across the carpet to a long conference table where a group of men and women lounged smoking cigars and drinking liquor from crystal highball glasses.

Troutman stood and said, "well ladies and gentlemen, here he is, the man we have been waiting for." Instinctively, McGill's red flags were up, and he was on high alert. This was entirely unorthodox. He was an enlisted man in the Air Force. A rank soldier among these politicians and career officers. He smelled a setup and waited for the shoe to drop. Every man and women in the room sensed his caution, for they had all seen it before. Hell, half of the room had been through it.

"Master Sergeant McGill," said a portly woman from the end of the conference table with absolute authority. The room immediately hushed, and the woman stood with some discomfort, dropping her cigar into the black ashtray, and reaching for the black cane that her aide offered her. "I am the Turtle," she said with authority. Everyone watched to see if McGill would smile, which he did not. This was good. "I have read your file," said the Turtle as she slowly advanced. "We have a serious situation that requires a very specific set of skills," she continued, hesitating for dramatic effect.

"My skills?" said McGill in what would only be described as an overly dramatic response.

"Your skills," said the Turtle. Now McGill did smile.

"You have been promoted to the rank of Captain and dismissed from the Air Force, your honorable discharge has been accepted and you will receive your full pension at this rank. From this moment forward you are assigned to Alpha as a civilian contractor. Do you understand?"

"Yes," was his reply. He had many questions, but this moment was not the time to ask them.

"Mr. Troutman, will you please provide Mr. McGill with a glass?" Troutman offered a chilled glass of amber liquid and a dark maduro cigar to McGill. The Turtle raised her glass and saluted McGill, leading the others to do the same. Confused and speechless, McGill raised his glass and downed it in a single gulp. It was the smoothest burn he had ever experienced.

The assembled men and women stepped forward to offer their hands, congratulate him, and slap him on the back. He was now one of them. Who "them" were, he still did not know.

The official briefing was done aboard the Puma. Troutman, McGill, and the smartly dressed Lieutenant Comden were flying Northwest. Troutman briefed him that the truck had been found, belonging to one Tobias Burlando, age 55, no occupation but locals described him as a prospector, which made sense given his location at the time of the crater. Forensic teams had been combing his home and found a trove of articles that had been burned or singed. Apparently, their visitor was hot. This would allow them to effectively use thermal imaging in finding it. Satellites were running over the area and they were now looking for an early 1980s Oldsmobile. While the focus was on interstates and highways, McGill's mind immediately went to mines and remote places that a prospector would have occasion to explore. McGill leaned to the Lieutenant and told her to provide reports for all mines in the area, active and abandoned. He returned his attention to Troutman, "do we know what he was prospecting for?"

"I thought all prospectors looked for gold?"

"No," replied McGill, "the area is known for hundreds of minerals and elements. Gold, silver, gems, uranium, tungsten. There are thousands of mines with miners looking for anything they can sell."

"Well, that complicates things," offered Troutman.

McGill leaned back to the Lieutenant. "Find out what he was mining for. Perhaps that will lead us to his most likely prospecting places."

While the Lieutenant rapidly emailed Command with her requests, Troutman and McGill continued reviewing files. Within the hour they had touched down at a small muni airport in Kernville where a black suburban was waiting for them. Within minutes they were in Kernville at the home of Tobias Burlando. As they walked through the house, they could smell the burned linins, wool, and leather. The chairs were singed in a perfect apple bottom shape. Troutman and McGill exchanged knowing glances and the Lieutenant rolled her eyes. Two perfectly formed footprints were charred into the hardwood floor in front of the fireplace. The scientist had done heat analysis projections, trying to discern just how hot it was. There was no radiation signature, no combustible activity other than what the suspect created, or other signs of toxic elements. It seemed to be pure heat emanating from an Alien being that left only hydrogen and helium behind. The lab would have a more comprehensive analysis in a few days.

"It is obviously awake and sentient," said Troutman. "Why is he taking it, or is he a hostage?"

McGill looked at the landscape of the room. He observed the tea and crackers. The comforters and the position of the chair next to the fire. "He is not a hostage. He is caring for it. He has befriended it, and it him, for reasons unknown. They are in this together."

"How far can they get in an Oldsmobile," Troutman asked rhetorically. He has one debit card and only a few hundred bucks in the bank...oh, by the way, he purchased a $50 wig when he returned."

McGill's eyebrows raised. "The old license plates on the Old's don't help us as our satellites do not ping on them and neither do the new highway patrol cameras. We will be searching the old-fashioned way. With visual and old-school hunting."

"What about the heat signature of the girl?" asked the Lieutenant, then catching herself and knowing her role was to be seen and not heard.

"We will only get a ping on that if she overloads. Unless she stands out like a beacon, we won't see her."

The Lieutenant received a call and listened intently before hanging up. "We just received a ping," she said.

†††

The suburban rolled to the shoulder of the road, then

towards a chain link fence off the highway. The "high volt-age" signs were wired to the fence and a group of troopers were milling around. The lieutenant immediately asked for the officer in charge, then rounded up the officers and escorted them out of the area. She would personally debrief and issue the non-disclosure threats. Troutman used a pen to observe the lock that dangled from the hasp. No doubt the local police had removed the lock and destroyed any finger-print evidence. McGill whistled low and drew Troutman's attention. He was looking at a tear in the fence where it had been melted, the aluminum drips rigid where they cooled. They followed the visual beyond the fence at the hillside and saw the charred line that imitated the sweeping arc of the fence melt. Last, they saw the melted rock. A bubbled, mass of hardened glass. "What the hell?" asked Troutman.

McGill turned his attention to the high voltage equip-ment, looking for clues. He found them everywhere. While a true technician has the correct tools and would hardly scratch the paint of this equipment, there were signs of a screwdriver having stripped out or wrestled with screws all over the device. He touched nothing, only made a mental note to have a utility chief out here right away. "Looks like they found what they wanted," said McGill. "Heat energy, lots, and lots of heat energy. We are now looking for spikes in electrical load. Anywhere that she can draw massive amounts of power. If I am right, there will be a spike when she con-nected here and charged up. If so, she will create a spike in the grid wherever she is."

"That's pretty efficient thinking," said Troutman. "I'm glad you're on our side."

McGill held his tongue but wondered what side he was really on.

The repair barn was cold. Concrete floors radiated the cold through the structure. Tobias and Red used the side door facing away from the parking lot and trudged to the dormitory. In season, fifty firefighters would live here comfortably. It was of new construction and had steel shutters rolled down over the windows to keep the bears and vandals from breaking in. The doors were made of steel, making them impenetrable to any thief hoping to pick a lock. Tobias looked at Red and gave her a "what if" look. She wrapped a small hand around the doorknob, and it melted in her palm. Still, the lock held, so she inserted her fingers into the space between the door and the door frame, melting both like butter, and slid her hand down, melting off the deadbolt. Tobias and Red stepped inside the dark room. With the windows sealed shut, it was pitch black inside. Tobias propped the still smoking door open, allowing light to shoot down the long hallway. Doors lined both sides of the hall, each with a cozy dorm room with two beds. While Red searched for the device that cranked open the windows, Tobias tried the gas heater and lights with no success. He walked out the back door and up the hill to the 2500-gal propane tank. It registered 3/4 full. He turned on the gas at the tank and returned to the dorm, finding the gas shutoff at floor level by the entrance. Soon he had each dorm room open wide and twenty-six room heaters turned on high. Red had cracked open the shutters on all the windows that were out of sight of the driveway, letting a beautiful sunlight flood the rooms and hall. While Tobias surveyed the large wooden roof beams and linoleum floors, Red sat on a gas radiator. Gas heat was akin to lukewarm water to her, but her entire body was cooling, and she needed to restore her heat.

To use the electrical, they would need to turn on the generator, and with daylight still out, there was no guarantee a visitor would not come onto the property. They would wait.

Each dorm was stocked with linens and bedding, carefully stowed to keep mice and insects out. Tobias made two beds in the room closest to the exit and brought in a box of food. As dusk came on, Tobias and Red walked in the nearby woods. She sensed animals all around them and as she sent these signals to Tobias, he described what they were. In the distance there was a small herd of deer taking advantage of the last portions of fall before they headed to lower elevations for winter. A bear was foraging around trash cans at the far end of the camp. They were empty, but he could smell the grease and dried jam on them. Too many birds and squirrels to count, each stocking away food for a long, cold winter.

When night was nearly on them, they walked back by way of the generator shed and Red melted their way into the room. Tobias was familiar with Generators as a mining staple and in just a few minutes they had the 365-Kilowatt generator running. Once the door was shut behind them, the generator made hardly a sound in the forest.

Tobias walked around the dormitory looking for an electric panel, which he found. Using his small toolkit, he removed the setscrew and lifted the panel, exposing two neat rows of breaker-switches. All were switched ON except for the main breaker. Using the utmost caution, Tobias switched all the breakers to OFF before switching on the Main. One by one, he switched on the individual breakers and had Red check to see that they were working. Red asked if she could connect directly to the system so she could warm up, but Tobias told her that everything in the dormitory was only 110 amp. She would need to charge directly from the generator for maximum efficiency.

Within two hours of arriving, Tobias and Red had brought in their supplies and created a comfortable and secure shelter. They had heat, food, and cover. Two hours after dusk, Tobias and Red donned their warm evening clothing and set off walking to the Generator house. They slipped into the noisy room and Tobias removed the plate covering the battery bank with a small toolkit in the room. They generator put 100% of its power into this bank of batteries, which were then used to regulate the power to the dormitory. With his thoughts, Tobias let Red know that this power source was free-standing, so she could take all the power she wanted without alarming the outside world. There were no sparks as on their first venture to a high-voltage line. She was drawing from 110 amps, which was like sipping from a straw. "It's too slow," Red said. Tobias looked over her left shoulder, his body close to hers. He could feel the static electricity stand his hair on end and reminded himself not to touch her while she was charging. As luck would have it, there was a 3-phase circuit, which is essentially 4 times the power of the standard household electrical plug. "Much better," she said, and she drew a reasonable amount of energy over the next hour to satisfy her for a few days.

Tobias shut off the generator and replaced the panel cover securing the panel. He shut off his flashlight and they walked back to the dormitory.

†††

Tobias woke with the uneasiness felt when someone is watching you. He opened his eyes to the dim light in the room. He had covers pulled snug up to his chin to ward off the cold. Eyes adjusting, he looked at the bunk next to his and saw Red staring straight at him. Her colorful eyes backlit as thought by a small bulb. Her head was hooded by her covers, but he could see the red sheen of her skin. She looked energized and healthy, he thought. He had wondered how much the cool night air at this elevation would affect her, but she seemed to be holding up ok. He sent a tendril of thought to her, "good morning."

"Good morning," she relied without speaking.

"Have you been awake long?" he asked.

"I do not know awake or sleep," she said. "I know move or rest."

He smiled. "How long have you been watching me?"

"All night," was her reply. "I have been connected to you,

learning what you know and sharing what I know. We have been exchanging information while you slept."

Tobias furrowed his brow and consciously thought of what she might know. He was surprised to discover the images that came to his mind. Images of the sun's brilliance, of the Children floating to and from their lives. Of the Lord and the Legions. Of Mer and Zel. A memory came to the front, it was of Red taking the life of a Child of the Sun. It showed the Lords great anger, the trial, judgement, and her ejection from the atmosphere of the Sun. Tears formed as he felt the intense unrelenting cold of space. It seemed he drifted forever, frozen in space and time. The sweet relief of the friction heat of entering Earth's atmosphere. Such un-restrained delight! She had not even felt the impact.

He smiled at her memory of first opening her eyes and seeing the ugliest Alien imaginable. Tobias opened his eyes wide and looked at Red in surprise. Laughing he threw his pillow at her, and she ducked under her covers, playing the game. "Ugly?" he said out loud. "I'll have you know that on this planet I am seen by some as handsome."

Red poked her head out of the covers and with her non-emotional, blank face simply said, "on a planet of ugly people."

Tobias laughed and shook his head. He reached back inside his memories, following them to where she touched him to access his mind. It was there that she observed his first impression of her. As beautiful. Tobias embarrassed raised his eyes to hers.

"I offer my apologies for not seeing you as beautiful as you saw me at first sight, she said. I see you now."

†††

After a light breakfast, Tobias and Red decided to reconnoiter the area to determine their safety. A sweep told them many things. First, they cut boughs of cedar and swept all the footprints from the previous night walk. They were very exposed up here, and it was cold. The forest offered privacy and solitude, but the cost was too great. They needed the warmth and electrical opportunity of the desert. They needed to head to a lower elevation.

They loaded the Olds in the early morning hours, replaced the doors in a reasonable condition and headed down the hill. At Kennedy Meadows, they turned right and continued down into the desert floor. High above Highway 395, they parked under a large pinyón pine tree and watched the traffic. Several black suburban's patrolled the Highway, along with the usual sheriff and highway Patrol units. Helicopters were running a grid on the far side of the valley and Tobias hoped they had already checked the area they were parked. Tobias was concerned and he fished his map from the glove compartment.

Tobias' knowledge of the Mojave Desert was encyclopedic. He had walked thousands of square miles prospecting for exotic minerals. There was much to be found here and Tobias felt he had only tapped a very small fraction of what the desert had to offer him. He knew he could not travel South. That was where he found Red. It would be the heaviest patrolled. He looked North and stopped almost immediately. How could he be so forgetful. He quickly folded the map and slid behind the wheel of the big car. He looked across the landscape below. There were dirt roads on the West side of the highway leading to where he wanted to go, but driving them would raise a cloud of dust, essentially signaling their location. Besides, his big car driving across a barren desert landscape slightly elevated from the freeway would stand out. They would be spotted in a moment. His only chance was to drop out of the foothills, turn on the main highway and make a run the eight miles to cinder road.

Feeling very exposed, he waited under cover of the tree until dusk. The moon was already over the mountain behind them, so no reflection of his vehicle would be seen from below. It was light enough for him to see the road in front of him, but dark enough that the Oldsmobile would not be seen from the highway below. They drove, lights off, until they reached the highway. When there were no headlights to the left or right, Tobias gunned the engine, racing up the highway, hoping to not be seen.

At Cinder road, he made a hard-right turn, heading East. Tires squealed and the old car rolled on her ancient shocks as he barreled down the pavement until it petered out to dirt. He shut off his lights again and drove slowly, careful to minimize any dust raising from their passage. Twenty minutes later, they could not see the road in the waning light. This was exactly where you would be if you were in 'the middle of nowhere.' The flat of the desert had opened into narrow twisting canyons. Tobias pulled on to the shoulder of the dirt road, shut off the engine.

Red felt his mind, understanding that it was too dark to drive and too dangerous to put on headlights in the desert. They would be seen for miles. They stepped from the car and stretched. The temperature was a perfect 75 for Tobias, and Red pulled her jacket close around her. The starry sky was riddled with stars. They were 100 miles from anywhere and the stars shone beautifully. Tobias smiled at Red and said, "beautiful, isn't it?"

Red sent Tobias a memory of an eternal floating in space with these stars as your only view. "No," she replied. "I don't find it beautiful at all."

They walked down the dirt road until Tobias could smell the sulfur in the air. He smiled; they were close. Another twenty minutes and they were at a tall chin-link fence. "After you," said Tobias. Red melted a large circle with her hand, and they stepped through the fence and walked in the starlight for another 10 minutes. The ground was riddled with silver pipes, each hot to the touch. They followed the pipes to the edge of a pond. The smell of sulfur was overwhelming. "Care to test the temperature?" said Tobias.

Red looked at Tobias and thought better of reading his intention. She kneeled and put a finger to the scalding hot water. She looked at Tobias with genuine surprise before stripping off her clothing and slipping into the water.

Tobias had been prospecting here many years ago. Twice he had been run off the property. The Coso Hydrothermal Facility was a series of deep-water pools heated from the interior of the earth. At 622 degrees Fahrenheit, they create electrical power through steam. The water would scald the skin right off a human, but it was the caustic chemicals that really killed you fast. To fall into one of these pools was a quick, painful death.

Red dove slowly, unsure of the depth. She found herself in narrow vents that continued downward. She converted her physical body into a gas form, and the heat instantly saturated every cell of her body. In the shallow pool was a pipe that brought this water up from the depths of the Earth. She flew through the water currents like a thicker, slower version of her flight on the Sun. The deeper she went the hotter it got, until she was actually warm.

"Tobias, this is amazing, go back to the car, I will find you when I surface."

Tobias, smiled, kicked a rock into the pool and turned on his heel. Tobias was never more comfortable than when walking in the desert starlight. Nothing but the sand crunching under his worn boots. The desert breeze playing over his skin. He was wonderfully happy, and he could feel Red's happiness as she raced through the Earth's crust.

Tobias woke as the sky was lightening. In the canyon, it would likely be 15-20 minutes before the sun was up. He brought his seat to an upright position, turned the key in the ignition and he backed the car deep in a ravine. Here in the desert, the rain does not soak into the dirt, it runs off. A small downpour of rain miles away could create a flashflood of water in a ravine like this. Over the course of time, it carved this gully, which was a perfect place to minimize visibility from the main dirt road. He rolled up the windows, got out of the car and slid his leather gloves onto his hands. For the next fifteen minutes he threw handfuls of sand over the top of the car, hoping to effectively camouflage it from the sky.

Tobias strapped his SOG seal knife onto his hip and he grabbed his short pick, a 2-foot-long handle on a small mattock. He cut a branch of creosote bush and wiped all footprints from the area, then hopped rock to rock up the slope to the ridge. There, in a clump of granite boulders, he could sit unnoticed from land or sky and survey the land for miles around.

Tobias sent out a morning greeting to Red, and she sent

back an image of her warm and comfortable, deep in the Earth.

Tobias was not concerned with the jets or helicopters overhead. He was turned towards the South, which allowed him to watch the roads and valleys to both East and West. He had moved several small boulders to make a larger space to sit and he found small common Opals. Soon, he was digging in earnest, unearthing a pocket of opal jasper of beautiful, brecciated color. He was finding mostly green and orange opal of moderate quality, but there were signs of quality in some samples. Occasionally he would pop his head up and look around, then return to his work. When he hit the layer of spotch, a black, crusty layer often associated with opals, he slowed his relentless digging and began searching. He did not hear the vehicle until it was close. He carefully popped up his head and looked down the road they had driven. There was nothing coming from that direction. He turned to the North and saw a black suburban traveling down the main entrance road, a quarter mile distant, to the hydrothermal facility from Coso Junction. He kept his head low and watched as the vehicle came to the main gate and then drove through.

"Red!" He sent her an image and she watched through his own eyes as he tracked the vehicle.

"Which pool am I in?" she asked.

Tobias glanced over a dozen pools spread over five square miles and stopped at one about a mile south of the visitors. They watched as the vehicle turned South and began an inspection of each pool. The driver never left the car, he just drove to each gate and gave it a visual once-over. Obviously, they were not suspecting anything. As Red raced to the surface, Tobias continued to follow the progress of the suburban. The web of roads leading to each pool and the network of pipes and controls associated with each made the search slow. Red was moving at a staggering speed, but she was still deep in the Earth when the driver came to the gate to the pool she was in. He parked there for about the same amount of time as the other stops, then backed out, turned, and continued his search, oblivious to the large hole in the fence one hundred meters down from the gate.

Tobias breathed a sigh of relief. And settled down to watch from a more comfortable position. When the search was complete, Red had reached the surface. She stayed underwater, watching through Tobias' eyes as the suburban drove back towards the entrance. It stopped at an intersection in the dirt road and made a left turn. Tobias was suddenly alarmed. This was the road he was parked on. The driver was apparently driving the full loop of roads in and out of the facility. The car stopped at the exit gate and waited for the main gate operator to open if remotely. He then drove through the gate, driving into the series of canyons that hid Tobias and the Oldsmobile.

Tobias lost the vehicle in the canyon below but could see the progress by the dust trail that rose on the winds. The suburban was traveling at high speed and blasted past the small ravine to his right. He never looked left or right, his eyes were on the road, and he never saw the Oldsmobile parked deep in the cut.

"Way too close," said Tobias.

They were sitting ducks in the desert. Tobias never gave much thought to how exposed you are when you have no place to hide. He looked from his perch at the wide-open landscape wondering how he would get Red out of here.

Red arrived seconds later, oblivious to her radiant red nakedness, but loving how strong and powerful she felt. The heat of this planet was far too little, but the atmosphere and oxygen supercharged her senses and her body. Perhaps the centuries she spent floating frozen in space had acclimated her to a cooler climate. Regardless, she was surviving here and for that she was grateful. With just a brief thought of the coldness of space in contrast to this moment by the thermal pool, she felt amazing.

Tobias, no longer shocked at Red's body, took it all in. The soft feminine curves, glowing red in the morning desert air. She descended to the car where she donned her suit and clothing. Tobias looked at the gas gauge, then at the map. Red could feel his concern, but in her current state of happiness, she could not join in his anxiety.

"Listen, I can feel that this water was great for you. I think you should stay here while I go back home and answer their questions. They might detain me for a few days, but they can't really do anything to me, right? I have not broken any laws or hurt anyone. What do you think?

"I only think what you think. All my knowledge of your world comes from you. You do not trust that your government will be honorable. You think they may hold you for much longer than a few days. You are afraid. For these reasons, I cannot allow you to go."

"We may have enough fuel to get to Coso Junction, but I fear they will be watching there. Certainly, the employees have been instructed to watch for our vehicle."

"Then we will leave the car here and travel by air."

"Red, there are no airports for miles around here."

"Who said anything about airports or airplanes," was her smug reply.

They experimented with flying together. Red had Tobias stand tall and she climbed onto his back in piggy-back fashion wrapping her arms and legs snugly around him. Tobias sent a warning, and she rapidly cooled her body between the two of them. She would need to be aware of managing her body temperatures. She lifted off, carrying him below her, and raced at high speed down the desert valley towards Ridgecrest. After a few miles she stopped and they discussed her available energy for a long flight and she was confident that it would be no problem, however, it was difficult for her to hold a dangling Tobias over long passages. They determined that if her were riding on her back it would be easier for her, leaving her arms free. On the flight back, Tobias wrapped his arms over her shoulders and leaned into her, belly to back. She glided forward into a prone position, and he settled into her, painfully aware of the curve and warmth of her body against his. Red was aware too and she sent a tendril to explore the feelings that Tobias was having. She found his reaction fascinating, wishing she could have them as well, but these emotions had long been bred out of her kind and she could only enjoy his pleasure.

Tobias wrapped his legs around Reds to keep from falling

forward and she leveled off to maintain their balance. Once they were comfortable, she increased her speed until Tobias held his head low and down to keep the wind from peeling him off. It was all he could do to hold on as they sped across the sky, Red slowed only enough to insure he stayed on, and they were back at the Oldsmobile in minutes.

Red left her clothing with Tobias and returned to the pool for one last dip in the water while Tobias ate a light breakfast. He donned his mining field vest, with its myriad pockets and he filled them with things that he did not want to leave behind.

Red headed for the pool, but feeling a draw of energy, she turned North, dropping into the fenced enclosure surrounding the power plant itself. She walked among the steam plants and electrical until she found the source she was looking for.

When Red returned, Tobias was on the ridge, pick in hand, carefully extracting stones. When he felt the heat at his back, he turned. Tobias was stunned. He had never seen her in a gas-state, and she was ablaze and rippling with heat. She looked like a desert mirage. "How did this happen," he asked.

"You just need to know where to look," she replied. "What did you find?"

Tobias held out a palm-sized stone that looked entirely normal except for the quarter-sized chip in the rough-brown surface, exposing the colorful, brilliant opal inside. "wouldn't you know it. I have searched a lifetime for these stones and now that I find them, I must leave them here."

Red easily lifted a few large boulders and covered his

excavation site. "There, no one will bother them now. We need to go." Red turned to the West and Tobias saw a trail of dust 20 miles distant. Cars were on the dirt road on the way to their location. To their South they could see the Helicopters as pinpricks on the horizon.

Red quickly donned her silver suit and Tobias wrapped his arms over her shoulders. She floated horizontally and allowed Tobias to arrange himself in a comfortable position, then she glided down the backside of the hill and flew East, into the desert.

†††

Due East of Coso is the Searles Valley. A desolate companion to Death Valley, which is only a few miles further East. Between the two valleys are the Panamint mountains. A raw, lava bed of barren wasteland. It is riddled with canyons and ravines that would erode swiftly if there were ever rain in the area. This was the driest, hottest, angriest place on earth. Red crossed the Searles Valley and climbed Panamint to Sentinel Peak. She circled once to see where the enemy was. There was no sign, meaning they were probably descending on the power station.

Red turned South and followed the rugged spine of the Panamint, dropping into the ravines and closed canyons when they were available. The Panamint mountains border the Eastern boundary of China Lake Naval Weapons Center. These were the same canyons Navy warplanes practice low-level flight and bombing missions. At the southern boundary of the Panamint range, she came out into open desert. She had left the Naval base and was now flying directly over Fort

Irwin, a Marine corps base. To her West was Edwards Air Force Base. She was literally in the belly of the beast. If they were spotted, they would be surrounded within minutes.

They continued for miles across the wasteland of the worst the Mojave Desert has to offer. The only dwellings here were rusted-out mobile homes and sub terrain dugout miners' shacks. Although outside of his normal prospecting territory, Tobias had walked much of this land, and he followed his landmarks.

Tobias, through squinting eyes, showed Red where to land and she swept into a hard-scrabble pile of rock Northeast of Barstow. Tobias was completely windblown. It was like sitting on top of a passenger jet for an hour straight. His eyes had long since teared up and dried out, leaving them crusty and stiff. His arms were quaking from holding on tight for the ride and his legs wobbled when he slid from Reds back. "Whew," was all he could say. "Some ride." He quickly ushered Red into the nearby mine, and they descended into the cool interior, out of the sun

"Is it safe to be here?" she asked.

"I want to make a stop nearby, but we need cover of darkness. There is a military depot a short distance from here. It is where all government equipment goes when it has outlived its usefulness. Miles and miles of perfectly good equipment baking in the desert sun. On occasion, I have visited auctions there. You would not believe what the military throws away."

Under the cover of night, they stepped out of the mine and into the moonlight. As they flew West, DRMO was clearly visible in the distance, its bright lights illuminating the desert landscape. DRMO is several hundred acres of surplus. Out in the open there are trucks, tanks, and tractors. Mountains of tires and batteries and gears and pallets of computers and processing equipment. There are thirteen massive storage building filled with everything from pencils to supercomputers. Red dropped inside the phalanx of warehouses where they would not be seen from the main base. Here it was dark. Despite its military use, there are no cameras or sensing devices at DRMO. In fact, none of the buildings are locked. All of this had outlived its useful life. To the government this was all trash and none of it mattered.

Tobias opened the door to the first warehouse, and they slipped inside. It was dark and quiet. Tobias fetched his flashlight from his vest, and they began searching. Within twenty minutes, the found a box labeled 'US Geologic Survey: Volcanology'. Inside were instruments and tools used in the study of volcanos, but no suits or clothing. They continued looking and it was three hours before they found another similar labeled box in the fourth warehouse they had inspected. Inside were plenty of suits, helmets, and tools, but none that fit any better than the one Red currently wore. When protecting from heat, you want something loose fitting. For Red to fly and move effectively, she needed snug fitting. While Tobias pulled equipment from the box, Red was searching the aisles and shelves. A box marked 'Vulcan Engineering Prototypes.' Caught her eye. It was a large steel cabinet locked by key. She melted the lock out and opened it. Inside were shelves lined with clothing prototypes for extreme heat. She pulled a metalized fluorescent green suit from a shelf and saw that it was clearly made for an exceptionally large man. She pulled several more suits out until she found one that suited her. It was form fitting and carbon black. It was snug, but not constricting, more of a wetsuit fit than a fire-suit fit. There were boots with Kevlar soles, mitts, which were totally useless for anything that required dexterity, and a full hood with face shield. Red put the hood on and sent out a blast of heat. The Pyrex lens immediately melted, as did boxes of equipment around her, but the cowl of the hood and her suit survived the blast. Within seconds, the fire sprinklers were on, and sirens were blasting across the base.

Tobias came running down the aisle as water poured from the ceiling, "what the hell happened?" He surveyed the smoking boxes and crates around her. He noted her stylish suit, but quickly rushed her out into the open where he climbed onto her back, and they flew east through the rows of warehouses and over the perimeter fence. She traveled a full mile before heading due South towards the San Jacinto mountains.

Red was visibly paler pink, but she looked nonetheless for wear as she walked up the gravel walkway to the broad deck and the large wooden door. Tobias knew exactly where they were.

When you live in a remote area with a locked gate three miles from your home, you don't expect visitors at 9pm. Joel was shocked awake by voices outside his house. He fished his revolver from the bed stand and quickly pulled on a pair of jeans. He peered out the full-length windows of his room and could see nothing in the inky blackness. He flipped on the floodlights outside and the entire area lit up. Anyone outside would be blinded. A knock came from the front door and he scurried through the master bed and into the living area. There he could see a disheveled, tired man and another person standing in the shadows behind him. "Who the hell is it?" demanded Joel.

"It's Tobias Burlando came the reply."

"Who?"

"Tobias Burlando, we worked a claim on your property a few years back."

"What the hell are you doing on my property?" spat Joel.

"It's a long story, but I need your help. Our truck broke down and we need a place to stay the night. Can we use the old caretaker's cabin?"

"What are you doing all the way out here at this time of night?" said the grumpy old man.

Tobias thought for a moment and said, "listen Joel, I know that this is a pretty weird way to show up. My apologies, it's a bit beyond my control. I will explain everything in the morning."

A moment's hesitation then the reply, "you know where it is. Come around tomorrow at 6am and we will square this out," said Joel, clearly agitated.

"Yessir," was the reply and the two retreated off the porch and walked up the drive to the old caretaker's house.

Joel turned on the driveway lights leading up towards the second gate where the caretaker used to live. He watched the two figures until they were out of sight before turning off the porch and lane lights. "Goddamned idiot," muttered Joel. "Ought to know better than to sneak up like that." Joel slept fitfully and finally made a pot of coffee around 4am. He paced his floor counting the minutes until 6am rolled around. He would not have to wait that long. A gentle knock on the front door at 5am and he grumbled," come in."

Tobias entered alone and crossed the room to shake Joel's hand. It did not look like Tobias had slept any better than he had. "Rough night?" said Joel.

"Rough night," repeated Tobias as he was passed a cup of scalding hot coffee. It was so black and strong that Tobias had to catch his breathe. He had forgotten how strong old Joel made his coffee.

"Got sugar?" asked Tobias. "Nah, I gave the stuff up, but I have some of that OHO raw honey. Damned fine in your cup. He passed a quart jar of the amber liquid and a teaspoon to Tobias, who spun a large rope on the spoon and dipped it in the cup. Stirring, Tobias offered is apologies for the rude arrival.

"Whatthehellyouthinkin?" asked Joel. "A man could get shot approaching like that."

"If you only knew," replied the other.

"Well," replied the old man, "I don't suppose you would do it without it bein' a mighty urgent thing. You wanna tell me about it?"

And Tobias did, from the beginning to the end while Joel was slack-jawed and frozen through the story. When Tobias finished, old Joel rose up on creaky knees and fetched the pot to heat up the now cold cups. He then brought the honey and spoon over and placed it in front of Tobias.

"Young man," he said. "You're about the most sober man I ever knew. I never knew you to be anything but honest, and of damned fine judgment, but I gotta admit this is pretty far-fetched. You know I'm gonna need to see it to believe it."

Red had been waiting outside, listening in to both men's thoughts as they talked and speculated. She chose that moment to come in. Dressed in her new black suit, her red, shiny head framed the prettiest face Joel had ever seen, and the eyes! They were the most amazing gems he could imagine. Joel scrambled to his feet and extended his hand, "Miss

Red, I'm Joel Kerry. This here is my place and your welcome here as long as you need." He could not take his eyes off hers; they were the gemstones he had hunted his whole life for.

"Thank you," she replied without moving her mouth.

"You readin' my mind?" he said.

"Yes Mr. Kerry, I am."

"Well I'll be damned," he said. "Can I fetch your coffee?"

"No thank you Mr. Kerry. I would like to join you if I may."

Joel woke from his daydream and quickly pulled out a chair for her. He then skittered off to the kitchen to prepare a quick breakfast for them. Tobias offered that Red did not eat but he would appreciate an egg or two.

"You'll have four eggs, some of my potatoes and a slab of bacon," he replied. "You all must be starved." Speaking to Red, Joel blurted out "Ma'am, if'n you don't eat, how do you keep goin?"

Red replied that where she came from everything a soul needed was in the atmosphere around them. "On Earth," she replied, "I guess I feed on heat. I need heat to keep going."

"Really?" Joel said objectively, "well if that don't beat all. How are you gettin' your heat, with fire?"

"I am currently feeding on any heat source I can find Mr. Kerry. Fire, water, electricity. They warm me and create this unusual pink color."

"It looks mighty good on you," offered Joel.

Tobias apologized and said that they had found an old extension cord in the caretaker's house and stripped a few wires. She had been feeding all night by wrapping the wires around her fingers.

"Well I'll be damned," said Joel.

"We will gladly reimburse you for your expense," said Tobias meekly.

"Well son," said the old man, "If I know you, you ain't got a pot to piss in, excuse me ma'am," he said blushing. "I went all solar out here two years ago and that 'lecricity don't cost me a dime. Ma'am, you just use all you need."

Tobias was overjoyed. Since they were not on the grid, they could not be traced. "How much energy do you produce?" Tobias asked.

"More than I need. Does she need more that she got last night? Sorry ma'am, I didn't mean to talk like you wasn't there."

"No apologies necessary Mr. Kerry," replied Red.

"Now there you go again," complained Joel. You either gotta call me Joel or nothin' at all. Mr. Kerry is my father's name, I'm just Joel."

"Very well," she said. "I function best when I have a high-voltage run through me."

"How much is considered high?" was Joel's persistent reply.

Tobias looked at him with a grin, "more than you got old friend. She tapped right into the grid and liked to suck it dry."

"Well I'll be damned," said the old man.

THE CHASE

McGill and Troutman had their loafers on the ground before the skids settled, and they walked across the meadow to a waiting Humvee.

As they approached, a wiry thin operator stepped forward and saluted, "Captain Orsen," he said, "Seal Team Two."

"I'm Troutman and this is McGill. Status?"

"Better to show you than describe it. You'll want to see this," he said. As they drove up the mountain road, the Captain described the chase leading up to this discovery. He noted that the facility was shut down for the season and only a chance stop by a forest trail manager caught the break-in before winter set in.

They pulled into the Blackrock compound and drove straight to the back of the property, where several black suburban's were parked. Captain Orson walked them to the dormitory and showed them the melted-out door locks. Finger impressions were clear where the Alien had inserted her hand into the space between the door and the door frame. McGill was handed a flashlight and he combed the interior. Burlando and the Alien had their pick of any rooms, yet they shared one. They used separate beds. One was piled high with blankets from several rooms and when he looked under the pile, he could see that the sheets were pressed and singed brown in the shape of a body. The mattress itself showed signs of melting. They were insulating the Alien to keep its heat contained.

They found the generator room in much the same condition. A panel removed from the wiring and rubber insulation melted away from the main contacts. It was clear that the Alien needed electricity or heat to survive here. What manner of creature was it? Where were the best places to find massive electricity?

The investigation continued through the night and the agents were tired. They hitched a ride with an agency vehicle heading to China Lake. McGill discussed his hypotheses on electricity and heat with Troutman. If it needed heat, would cooling it capture, harm, or potentially kill it? Their orders were to capture. They needed to know exactly what they were dealing with before they inadvertently destroyed it. If it did not pose a risk to either public relations or population, they were not to harm it.

Troutman filled McGill in on his own investigations, explaining that based on the gas chromatogram readings from the crater, the science teams had adjusted the satellites to look for hydrogen and helium. It took them a while to calibrate, but they were able to discern a trail running from the crash site to Burlando's house, then up into the high-Country towards Blackrock. "They are tracking towards Blackrock now. It is a slow process and it is unknown if we can analyze the data faster than they can travel, but we now have a trail to follow."

As dawn arrived, Troutman and McGill were arriving at China Lake. Officers billeting on base were far more comfortable than a hotel. They would get a few hours' sleep before the next update would put them back on the road. The last call was from Troutman to the helicopter pilot to be ready to leave on a moment's notice.

There are a few sure things in this business. *Worst possible timing* was one of them. About the time the agency men got to sleep, the call came. Feet hit the floor, a little slower than before. Pulling on stiff new jeans, a stiff new shirt, and stiff new boots that Lt. Comden had delivered from the quartermaster's office. They looked like catalog models for REI. They walked out the door at the same time and the driver was already pulled up out front waiting for them. Hot coffee in the rear seat cup holders and a box of mini-mart doughnuts in the seat between the men, this driver was a keeper. The drive to the flight line was short and they were in the helicopter within minutes.

No sooner were they airborne, than they were descending into the fenced compound of the hydrothermal facility. Both men frustrated that they had been this close and not even knew it. The Alpha Agents ducked out of the chopper and ran through the stinging swirl of sand towards the main office of the facility. As they stepped into the office, they were greeted by a weathered hulk of a man wearing denim work pants, boots, and a white t-shirt. A pair of leather gloves jammed under the belt that boasted a buckle as big as a saucer proclaiming the wearer the winning 'Bishop Mule Days Bull Rider'. He looked like the Bull could ride him. Oliver Santos was the facility foreman. He had received a spike notice on his cell phone and drove out to inspect the systems

Santos led the agents and Captain Orson between rows of geothermal steam engines to a large steel box that had been warped and twisted by intense heat. On either side was a small handprint burned into the surface. "What do you make of this?" asked Orson.

"Nothing at all said Troutman, and neither does your team. This is now classified Top Secret. You will discuss this operation with no one. Your superiors will get a redacted version that omits these handprints, do you understand?"

"Yessir," was the reply. It was good to work with professionals.

"Make certain to secure the site and brief Mr. Santos, would you?" continued Troutman.

Yessir," was the reply.

"Mr. Santos, can you tell us how much power runs through this panel?"

"It fluctuates, but right now we are creating about three hundred mega-watts."

"That's a lot of juice," said Troutman.

Science and tactical teams were arriving by air, more would arrive by car. Their proximity to China Lake would speed this up. Within the hour the site was overrun with government men and women. As they got to work, information began to come in rapid-fire.

†††

McGill and Troutman got out of the car at hydro pool number six. The area was flagged off-limits and only two military personnel had entered the fence-line. They were well trained and did not disrupt the evidence in the area. Troutman and McGill walked in the same footprints as the previous team. They came to the fence and noted the same drips of cooled aluminum on the fence cut that had occurred at the transformer station in Kernville. They stepped through the fence and walked to the water's edge. Sulfur assaulted their noses as they peered into the aqua-blue depths.

Troutman noted that she may have fallen in and drowned. McGill was not so certain. There was no stumble, and the toes of each foot were clearly indented, like a jump. McGill walked the perimeter looking for any sign that the Alien had climbed out of the water. None at all. Two sets of footprints entering, only one leaving, and that was Burlando. Where was the Alien? "How hot is this pool?" he asked Santos.

"About six hundred degrees," said Santos. "We have 150 wells in the area, about thirteen have an open pool like this one. Most of the wells are two to three thousand meters deep, but the source that creates all this heat is two to three kilometers underground. It's a field of molten magma. Temperatures down there are over one thousand degrees."

Troutman and McGill exchanged knowing glances. It was after heat. The big question was, is it still down there?

†††

The Oldsmobile was found shortly thereafter. At the old car, McGill noted the cleverness of covering it with a dusting of desert sand. You would never see it from the air. They hiked up to the ridge and could see where someone had been prospecting, and the obvious movement of large boulders to cover the location. "Get a team to move these boulders," said Troutman. "I want to know what is under them."

McGill looked across the horizon in all directions. Burlando would be able to see anyone coming for miles around. He would have plenty of lead-time to escape, but on foot? In this desert no man can last more than a few hours without water, even one as experienced as Burlando. Besides there were no footprints other than the ones heading to the pool. There were none leading out of the area.

McGill mulled this over a few minutes before reaching for his phone. No signal so he reached into his go-bag and removed a sat phone. He needed a full grid pattern search with thermal imagery. He wanted every satellite possible pointed at one hundred square miles of desert. He wanted every person in open desert stopped and questioned. He wanted answers.

The agents returned to the plant offices and McGill wandered off to see what the science teams had discovered. He walked between massive steam engines to the panel box the Alien had accessed where he watched the forensic men work. They had some new devices working that he had never seen, and an oscilloscope of sorts reading the area.

"Report," said McGill.

The lead scientist was exactly what you would expect. Thick glasses, tousled hair and a breast pocket stuffed with a scientific calculator. "Our clearance only allows observation sir. We have been asked to report back the elements found on this site. So far, the only unusual elements are elevated levels of hydrogen and helium with trace amounts of cobalt, carbon, neon, and iron. Pretty odd sir. I'm not aware of anything that has this particular elemental signature short of the sun."

"The sun-sun?" asked McGill. "The big ball in the sky?"

"Yes, said the scientist. 73% hydrogen, 25% helium and trace amounts of oxygen, cobalt, neon, and iron. That's what that big fireball in the sky is." The scientist chuckled at his little joke and let it die as his audience did not join in his levity.

"Keep us posted," said McGill, then he was on his phone.

At the agency headquarters, tempers were wearing thin. The science team had been working nonstop for days with terrific breakthroughs every hour. They had isolated the elements at the crater, they had discovered the hydrogen trail and their satellites and computers were working overtime to compile data on the hydrogen trail. Each square on the computer screen represented ten feet and the screen moved one square every one to two minutes. At this rate, their prey could walk across country and still not be caught. With all this combined knowledge, computer, and brain power, particularly when their subject had fallen from the sky, there was not a single member of their team who was knowledge-able about space.

Troutman's call corrected that, and many egos were bruised that a simple field tech had solved a riddle they missed. Looking for a single element was a needle in a haystack. The more data you had, the quicker the system would work. The analyst entered the precise elemental mix of the suns atmosphere and the movement sped up appreciably, to one square every 30 seconds. Not a blistering pace, but a definite improvement.

When asked why she chose to stop at Joel's Ranch, Red replied that the answer was in Tobias' thoughts. For years, Joel had been mining tourmalines on his property. Tourmalines are a tremendous source for storing energy, she noted. Many crystals can achieve this through pressure, but if your thoughts on the matter are correct, we should be able to store vast amounts of energy in them by compacting the energy inside. "I don't understand," said Tobias.

Joel caught on immediately, "so your sayin that if you jam enough energy into a tourmaline stone, it can pack in densely and store that energy to use later, kind of like a super-battery?"

"Yes," she said.

"How big does this stone need to be?" he asked.

"I don't know until I try it," she replied.

"Well I'll be damned," Joel said as he retreated to his master bedroom and returned to the porch with an assortment of bi-color and watermelon tourmalines of various sizes.

Tobias was dumbfounded. "These are worth a bloody mint," he said.

"I've been saving them for a rainy day," said Joel.

Red picked up a medium sized bi-color. It was about 80 grams, which is large for any kind of gemstone. She walked out into the driveway, which was made of small, pebbled granite. She hefted the stone in a closed palm and routed energy into it. The stone took what she offered without problem, so she sent more. Still fine. She sent a burst of energy to the stone and it shattered into shards, riddling the men with sharp little missiles.

"You could put an eye out with that little lady," said Joel.

"How much energy did you put into it?" said Tobias.

"A lot," replied Red. "Perhaps twenty to thirty percent of my capacity."

"So, a dozen of these could recharge you?"

"A dozen could recharge me very well," was her reply.

Joel looked at the red dust covering his porch. "Well," he noted. "That was probably a $20,000 stone you obliterated there. Can you control it?"

"I'm sorry," said Red. I could feel it overcharging but I needed to know how far it could go. I think I can make it up to you."

Noon found the three of them in Joel's pickup truck driving to his solar farm where Red topped off her reserves, while draining his. Then they headed to the mine-site. Most gems mined in California come from an underground layer called pegmatite, or 'peg' as miners call it. Think of it as the birthplace of gemstones, containing all the properties necessary to create these crystals. Throughout the Baja Peninsula pegmatite is buried 5 or more miles below the surface, but when it comes into Southern California, heat and pressure under the earth lifted mountains, bringing the up pegmatite with it. As the mountains eroded away, they exposed this pegmatite and the gems found within. The finest gems were found inside of pockets or "vug's," abscesses in solid granite where they have room to grow. It is not easy to mine tourmalines from solid rock. Joel had found large pockets filled with treasure, only to find nothing for two or three years thereafter, then hitting another rich pocket. It was feast or famine. He was used to setting aside gems for the lean times.

When they got to the mine, Red held her hands palm out and turned in a slow circle, searching for energy. The weight of the earth on the gemstones created an electrical pressure. She was hoping to feel it, but at the mine site this feeling was very weak. They walked around for a half hour with no success. In a last-ditch effort, Red lay down belly first in the dirt, arms and legs outstretched and sent out a single powerful burst into the soil. No one breathed while they waited for a response. Red stood up and brushed off. She then started hiking to the East, up the slope of the valley. She sent a message to Joel to bring the dozer.

Joel returned to the barn, fired up his D8 bulldozer and crawled it to the slope Red was on. From here, she sent him a mental image of where and how deep. He began cutting away the hillside one long cut at a time. Each cut exposing a long dike of pegmatite. Two hours later and he had Tobias return to his truck to fetch some diesel fuel and oil. While Joel refilled the diesel in the tank and serviced the dozer, Tobias and

Red walked across the cut. The dike had been exposed perhaps six-foot-tall and thirty feet long. The dozer was no good at cutting granite. They would need the jackhammer. True gem miners did not like blasting in the peg as it could damage a rare specimen stone. "Here," she said, pointing to a section of solid granite.

"How deep," asked Tobias?

"Inches," she said.

With that, Tobias fetched a gas-powered jack hammer from the truck. It was heavy and cumbersome, but it would eat away at the peg. He fired it up on the first pull and set the iron point at the location Red had noted. He chipped away for thirty minutes before the tip sunk into a hole. Within 20 minutes more he had widened the hole large enough see inside. On hands and knees, using the powerful flashlight, Tobias peered into his fantasy. A massive pocket, nearly a cavern of tourmaline clear and pure, in an area they had never mined and likely never would have. There was no indication at all that the peg ran through here. Certainly, no sign that anything this remarkable was here. He came to his knees and called to Joel, who was cleaning his hands of the fuel and oils. Tobias passed him the flashlight and Joel Kerry, lifetime gem miner and mineral enthusiast kneeled down to the culmination of his life's work. He reached inside the hole and pulled out a blue-cap tourmaline, clear as glass and over 2000 carats. It was a floater, terminated on both ends, meaning it was a pristine specimen. The first piece out of the pocket and it was remarkable. What else would this hole produce? Joel looked at Red and, with tears in his eyes asked, "how did you find it?"

Red's response would be hilarious if it were not so practical. "You have hundreds of pockets in this valley," she said. "You could not unearth them all in five lifetimes. You follow the pegmatite you find, but it rises and falls all around you. I simply looked for the most powerful and resonant source that could quickly be unearthed. This large cavern seemed like the best place to start."

Joel looked around his valley. He had earned a hard-scrabble life for many years. He had socked enough away for this little piece of land and he had worked it diligently, creating a life for himself, but never understanding or knowing security. Today he understood it.

Over the course of the day, they opened the pocket further, allowing them to lower themselves into the hole in a harness and work with care on removing the tourmaline specimens. The most valuable were the clusters, which they left intact for Joel to remove at his leisure. He would take days or weeks to painstakingly remove each cluster with pointed wooden skewers and small hand tools, careful to keep the fragile mineral group intact. Floaters were single tourmaline crystals that grew individually in the clay-rich pocket. They removed about forty floaters upon opening the cave-like structure in the pegmatite. Red had hefted each of these in her hand and selected twelve that were each roughly the size of a roll of quarters. She claimed these as her fee for finding the pocket, and they left the rest to Joel Kerry. In addition, Red made small cairns of stone in a dozen places that Joel should excavate. She did not do more than mark the locations because she did not want to spoil his fun of finding the pockets. For his part, Joel was 20 years younger as he gently removed each gem from its hiding place and reverently handing them to Tobias, who cleaned them in water, lightly brushed off the dirt and then handed them to Red to evaluate.

Red spent the rest of the afternoon alone at the solar battery shed. She would funnel the energy from the panel directly into each gem, carefully compacting it inside. When it was as full as she was comfortable with, she would slide each into an ammo belt designed for shotgun shells that Joel had in his barn. She could wear the belt and retrieve a stone in seconds, ready to recharge or warm her. It was a good day. One where her energy was strong and with the help of the carbon suit and extra clothing, she remained moderately comfortable.

Tobias and Joel remained at the pocket, too excited at the find, and the joy of unearthing each and every stone. When the light of day had faded to the point it was difficult to properly work, they reluctantly packed the truck and headed to the solar shed to fetch Red. They stood by in awe as she packed the last of the stones, sliding it into a slot on the belt.

While Joel drove his truck back to the ranch house, Red and Tobias walked across the ranch property in the dusk, enjoying the quiet serenity. The bullfrogs were croaking at the pond and the red-wing blackbirds were swarming the night insects. Their song was strongest in early morning and early evening. A doe and her fawn raised their heads from drinking and watched as the two passed by. Red could sense all manner of wildlife around them and she shared her perceptions with Tobias, who had been oblivious to the bobcat and the horned owl who were nearby. Both hunting the many mice and squirrels that were out foraging.

When they arrived at the main house, Joel had already washed up and was fixing a peasant's feast of salami, cheeses, and fresh sourdough bread. He poured two water glasses full of a homemade elderberry wine, made the year prior, and Red sat contentedly while the men ate and told stories of their day.

That night, Tobias woke while the moon was high and walked barefoot out on the veranda. In the moonlight he could see Red sitting cross-legged on the lawn next to the creek that ran through the property. He found a rocking chair and the wood creaked as he gently sat in it. She was aware he was there, but she did not turn. She felt the earth in her fingers. She watched the bobcats hunt and the cattle mill around the pasture beyond. A lone bullfrog croaked, waiting for an answer, and getting none. They were still sitting there as the stars faded, the sky lightened, and the day began.

Tobias rose and walked alone to Joel's cabin, enjoying breakfast and coffee with his good friend. When Red joined the men, she advised that she was going out to hunt for energy resources. She may be gone for hours, perhaps a day, but that she would return once she had a read on her abilities. Tobias was against it from the start. He did not like the idea of Red on her own without his help, but she reminded him that he was always with her. She could move farther and faster alone.

The men worked in the mine throughout the day while Red flew around the solitary valley. She was reaching far with her mind, connecting to the animals and the earth. She reached beyond to the local communities and then to the cities in the valleys far below. She searched for danger and found none. As she explored her own abilities, she discovered that her ability to move between her physical and vapor presence was completely negated by her suit and bandolier. The minute she shed her physical form, her suit and energy stones simply fell out of the sky and onto the ground. The challenges are always in the details, she thought.

Red returned at dusk and recharged in the barn before returning to the main house. Both Tobias and Joel were visibly relieved to see her and the talked excitedly through the early evening about her experiences. Around ten o'clock, Tobias and Joel turned in for the night and Red shed her suit and crystal bandolier, leaving it on the kitchen table. She stepped outside and walked to the pebble walkway. She flew low and slow down into the Temecula Valley, then continuing directly through Camp Pendleton Marine Base to the Pacific Ocean. Her objective was just North of the military compound on the border of Oceanside beach.

She touched down inside the towering gates of the San Onofre' Nuclear Power Plant. She walked directly to the giant transformers that collected the energy from the plant. She could feel the static charge peeling off the transformer. Just being in proximity to this energy was comforting. She could feel the massive power of fission occurring in the plant itself. A fission event can create over 2 trillion Kelvin, or 375 trillion Fahrenheit degrees. She wanted to go into the belly of the plant itself and bathe in that warmth, but the radioactive nature of a nuclear heat would be harmful to others she cared about. She must be satisfied with electrical heat. She wrapped a hand firmly around one cable, then placed her palm out toward a second, feeling the intensity and regulating the intake of energy. It was far more powerful than she thought, it was amazing. She moved her hand closer and her skin shone crimson. She was taking in energy heat at an astounding rate. Gulping it down like a man dying of thirst. She held her position for five full minutes, realizing that she still had much more capacity. This was too slow. She grabbed hold of each wire firmly and the lights of the city dimmed.

While Red was charging, the facility operators were frantically trying to ascertain the cause of a massive power draw. Computer screens held the attention of operators, engineers were double checking their systems. Like ants on a hill, they scurried about to find the source of the breach.

THE HELIUM TRAIL

McGill studied the map of the helium trail. By measuring the trails progress and decay, scientists estimated that they were moving cross-country at over 80 miles per hour. There was no disturbance of soil or canopy along the route, so it was determined that they must be levitating or flying. This changed everything. Later reports showed when they left the hydrothermal facility, the trail shot due South in a straight line at high speed, crossing both China Lake and Edwards bases. Neither the Navy nor Air Force registered any breach. This was now a major National Security issue. McGill imagined many heads rolling as Washington panicked. Yes, they were flying, and at an undetermined speed in a direction that could not be anticipated. No energy spikes had been observed since the hydrothermal facility, so it had stored enough heat energy, or it was drawing from untraceable sources or off the grid. Although technology was being employed to speed up the tracking, he did not see them catching up with the pair unless they stopped for a few days.

A rap on the door and Troutman's voice said, "wheels up in 10." McGill was packed and ready. He slung the black pack over his shoulder and was out the door and following

Troutman in seconds. The chopper did not follow the trail route, it cut directly across the desert towards Barstow, with two additional helicopters in escort. On the flight they observed that the crossing of military bases was bold, but actually breaking into a Marine facility was very risky. It must have needed something in there very much.

The tour of the DRMO base confirmed the conclusions that they were drawing. They had been searching for ways to minimize heat loss from the Alien. The missing suit confirmed they were looking for a way to minimize its heat loss, and the flash burns which melted the face shield and ignited the interior of the warehouse showed the tremendous force of its power. This could be an entirely new weapon for the US Military.

†††

The helium trail was slowly tracking the pair to Palm Springs. Alpha was updating the trail in real time to all agents and half were following the trail while the other half were anticipating movements by canvassing all territory out in front of the known trail. Satellites were actively scanning a one-hundred-mile radius around the trail and the advance team was checking every lead. Red had started erratically turning East and West after DRMO, but when the agency lined up the overall progress, it still gave them a fairly predictable path.

Each day, Troutman and McGill holed up in a different hotel, eating in different restaurants and they completely immersed in the hunt for the Alien. Their reports to The Turtle were hourly now. The supportive leader façade was long gone, and the Turtle had become a vicious, sniping taskmaster. Her anger was legendary and every hour she berated and dressed-down her agents for not moving faster and finding her Alien.

Any alarm at San Onofre Nuclear Power Plant is a very big deal. These are not managed on a local level like the hydrothermal facility. They are connected directly to Camp Pendleton Marine base, the Governor's office, and the Pentagon. From the moment Red breached the facility, the alarm was on. Within minutes, helicopters were in the air from only five miles out. Ground troops were loading into troop transports and local law enforcement was stopping all traffic on nearby Interstate 5.

McGill and Troutman were wheels up from Palm Springs Airport within minutes were on the way to San Onofre at top speed. While Red took hold of the wires and enjoyed the powerful surge of energy, the security teams at San Onofre were climbing the scaffolding to the second level panels. What they saw scared the hell out of them. Red was in a semi-gaseous state, hands firmly on the cables. She was brilliant crimson, but transparent enough that they could see the equipment on the other side of her. Sparks flowed from her hands and waves of heat drove security back.

Helicopters dropped teams via fast-rope. The freeway was a line of law enforcement lights. Troop trucks barreled through the gates. Red was oblivious to all of this as she fed deep and long of the electrical heat. When she let go of the cables, she was a fiery form. The steel cabinets around her were warping with the heat, the steel scaffold she hovered over had a hole melted directly under her. She looked through kaleidoscope eyes at the forms around her. Weapons trained on her as a helicopter approached with a large flexible hose hanging from its belly.

McGill and Troutman flew into visual range as the Marine helicopter began ejecting liquid nitrogen from the hose. Red simply glided twenty feet to the left as the nitrogen instantly froze the scaffolding where she stood seconds before. The intense cold hitting the molten heat, shattered the steel, showering the troops with steel shrapnel and driving them back. The helicopter adjusted, and Red simply moved out of its range each time.

Every eye on the Alien was transfixed. It was both frightening and beautiful. Since the moment McGill saw the cheeks burned into the chair in Burlando's home, he felt certain that his was a female form. To look upon her was to behold true feminine beauty, and true feminine intensity.

No one knew who fired the first shot, but once the gunshot was heard, a barrage of gunfire was set free. Everyone was tense and the friction was palpable in the air. When it was released, it was an all-out shooting gallery.

The bullets passed harmlessly through the gas form of Red, but she could see what her future held. She now understood why Tobias had fled rather than bring her to the authorities. There was not greeting, no welcome. There was only violence and death. It would be their death.

The first beam of energy struck the Marine Helicopter dumping the liquid nitrogen, disintegrating it in midair. Steel and carbon rained down as dust. Red turned her force to the troops, melting the weapons in their hands as they screamed in agony while their fingers melted against the white-hot gunstocks. The troop carriers were obliterated and blown into the ocean, some 600 yards away. Red rose in altitude until she was face to face with the helicopter carrying Troutman and McGill. She looked directly into McGill's eyes. He was entranced by the myriad of colors in hers. Then she was gone, flying due South to Mexico.

Two helicopters descended into the meadow opposite the ranch house and one circled above watching.

McGill, Troutman, and Comden exited the first helo and the second disgorged a dozen tactical troops in full gear, assault rifles at the ready. The tactical soldiers fanned out and made a quick search of the home and outbuildings. The third helicopter called in that there was a truck a quarter mile down the canyon and two older men watching them from the ground. McGill could hear the loudspeaker on the third helicopter telling the men to return to the ranch house, followed by a radio response from Helicopter 3 that the men gave them 'the bird'. Two tactical members of team-two set off at a fast jog down the canyon and shortly returned driving the old man's truck with both men sitting handcuffed in the back seat.

McGill and Troutman watched them approach from the deck of the main house. They were sitting in comfortable rocking chairs, waiting for the men to arrive. The soldiers gently ushered Joel and Tobias out of the truck and marched them up to the deck. McGill, rose from his chair and instructed the solders to remove the handcuffs, then he walked down the steps to stand in front them. He extended his hand and said "Tom McGill."

Neither man accepted his hand but stared impassively into the eyes of the Captain. McGill lowered his hand and smiled. He liked these old buzzards. He said, "do you have any coffee? I could sure use a cup. We left before I finished mine." Joel walked right around McGill and into the open door, gently shoving aside the soldier blocking his path. The old man certainly had sand. Tobias stood stonily in front of McGill. He was calm and at peace, like a man who knew exactly where he was in the world.

The soldiers had formed a perimeter 40 yards out, and McGill and Troutman sat on the porch with the two men, blowing on the hot coffee. McGill's face contorted with the first sip, not certain if the old man really liked his coffee this strong, or if he were making a point.

"A rope of that honey makes it a mite better," said Joel, pushing the jar of OHO raw honey towards them. Both Troutman and McGill put some in their cup, hoping that anything would take the edge off. To their surprise, it was quite good.

"Would you like to tell us about your visit?" said McGill.

Tobias blew on his coffee and acted as if nothing was said.

"This is a nice piece of property; do you ranch it?" asked Troutman of Joel.

"Nope, I mine it," said Joel.

"Is that how you two met?" asked McGill. "Are you old prospecting buddies?"

"My friend here is the salt of the Earth," said Joel. "He is as honest as the day is long and I trust him with my life. He can trust me with his life as well."

Tobias looked up from his lap to his friend, offering an appreciative smile, then returned to staring at his lap.

Both Troutman and McGill appreciated this answer. It spoke to the very character of friendship, honor, and loyalty. These were character traits that both men admired. They did not want this old man to betray his friendship, but they had a job to do. McGill said as much; "Mr. Kerry, I appreciate your loyalty and faith in Mr. Burlando, but you know that we must find his companion. You know what our job is."

Kerry sat silent for a moment and prepared a pipe full of tobacco. He lit is and sat back deep in the chair. "I do understand. I was a service man myself in a time past. I know what orders is, and I understand you have a job to do. I believe you when you say that you understand my loyalty. That man and his companion are now like my blood. I will watch over them and protect them with my dying breath. You have him here and you may find her eventually, but you will not have my help in doing it."

Troutman keyed in on his reference to the Aliens as 'her.' "Is she okay? How is she assimilating to our world? What are her needs? Does she have any demands or requests?"

Kerry just rocked deep in his chair, looking over the pasture and smoking his pipe. Tobias never said a word, afraid that he would betray his friend.

Red sat in a deep recess in the granite cliff high above the ranch, watching through the eyes of Tobias and Joel, and listened through their minds. She did not feel that they were in immediate danger and was patient in her wait. It was really no different than hunting the Damned. Patience would resolve everything.

For the next two days, the team stayed on at the ranch, commandeering the caretakers house as a barracks and flying in teams of scientists to comb the area. The soldiers set up a field camp in the pasture, displacing the horses and mules that were moved to a small corral. The scientists had travel trailers trucked in so they would have room to both work and sleep. Most of their best clues had come from the science community and they hoped to build on the knowledge that had been learned this far.

Each day, Joel and Tobias were questioned. Through constant cross-examination, they had given up much of what they knew without knowing they had given it up. Such was the process of interrogation. The big reveal was video of the attack at San Onofre. Both Tobias and Joel were stunned by the violence of Reds reaction. The reality of military men and women dying and grievously injured was clear. They would never stop pursuing her. She was a perfect weapon for a country seeking ultimate power. There was no middle ground here.

McGill treated both men with respect and care in an effort to win confidence. Each time McGill made a little headway, Troutman would bust in demanding answers and completely destroy the trust that was being built. Troutman was getting a great deal of pressure from the top and he needed results. McGill's tactic of soft interrogation was not getting them anywhere. Soon, Troutman would do it his way.

Eventually, the scientists would find the shards of exploded tourmaline, although they would not piece together how it happened or why. Joel was questioned about the cairns and he did not betray a word. Holes were dug at each and electronic samples were taken, but none revealed what Red knew and the men knew. A science director wanted to seize all the tourmaline samples, but McGill killed that abruptly and told the scientist to do his job.

As for the helium trail, that had been completely obscured by Reds circling and exploring the Valley and the surrounding area. The trail overlapped itself so many times that the scientists were having difficulty determining where it actually was.

On the third day, McGill left his card with Tobias, asking him to please let him know if Red returned. McGill and Troutman then headed back to Edwards, the science team had packed and left, and Joel and Tobias were held under house-arrest with a security detail. Red determined that it would be a good time to get answers and she left her perch and flew East.

The Turtle was furious. Under constant threat of replacement or defunding from above, she had made promises that she could not keep. They had the old man, hell, they had two old men. Both knew where that Alien was, but neither was talking. McGill had assured her he would manage it, but he had failed. Now she would assume command. She orchestrated the intelligence and data, issuing orders and putting pieces in place. The Turtle boarded her helicopter and they landed in the pasture less than hour after Troutman, McGill and the scientists had departed. As she limped across the pasture, she shouted orders to have the prisoners taken to the main house. Tobias and Joel, who had been sitting quietly on the porch were driven roughly to the deck, handcuffed, and prepared for interrogation.

The Turtle was in a foul mood and she berated the soldiers for their failure to find the Alien. She berated Joel and Tobias for their treason. She raised her cane and brought it harshly down on the neck of Joel, who crumpled under the blow. Tobias started to object, and the Turtle raised her cane again, threatening to attack Tobias as well. She ordered Tobias held while she beat Joel with her cane. There was not much strength in her, but that did not mean the blows did not have effect. She ordered Joel be sat at his kitchen table, a large slab of sugar pine milled and beautifully made right on this property.

She had Joel's hands nailed to the table, while Tobias howled in anguish at the inhumane torture being dealt to his friend. The Turtle turned to Tobias and through clenched teeth told him that this was his fault. His friend would be beaten, tortured, and perhaps killed, because Tobias was a traitor to his country.

As she had a fingernail removed from Joel's hand with pliers, Joel smiled through bleeding lips and said, "you will be in big trouble when she finds out what you did." The punch to the side of his face was his final conscious thought before he blacked out.

†††

Red was alight with flame and energy, flying over the Nevada desert, far to the East. She felt powerful and danger-ous. The level of energy she had consumed at the nuclear power plant had far surpassed any she had enjoyed since her arrival on this planet and it fueled a part of her Damned self that conflicted with the passive nature of Tobias, whom she had deeply imprinted on. She fought for control over her Alien side and won, calming herself to a point that she could think clearly. She reached out her mind to Tobias and found him shrieking and fighting with a soldier. She reached out to Joel as his pain-stricken thoughts said, "you will be in big trouble when she finds out what you did." She felt the punch to his face, and she felt him black out.

In her heighted energy and mental state, she did not care that she left a trail of flame in her wake. She saw only the rage that comes from someone you love being hurt. Her speed was tremendous, virtually instantaneous as she crossed the desert, climbed Mount San Jacinto, and shot across the plateau like a bullet. She landed next to the Puma helicopter at the Ranch and her trail of flame completely engulfed the vehicle and vaporized it as the heat whiplashed to the ground. The soldiers loitering in proximity the aircraft were roasted. Flames leapt up from the trees and brush for 200 feet in every direction. The paint peeled and burnt off the side of the house, but the structure did not ignite. Red walked to the driveway and called to those inside to come out. The Turtle smiled cruelly at Tobias and said, "looks like she came to me."

The turtle limped out the front door flanked by four large soldiers. They were seasoned veterans facing a young woman, albeit a flaming red one. They came out of the house, weapons leveled squarely at Red, ordering her to stand down. The Turtle has a look of supreme arrogance on her face.

The guns melted in the hands of the soldiers, the molten steel pouring over skin and bone, then hardening. The men stared at their hands, screaming in pain. She then shot a single bolt of heat through them, melting a hole clean through their middle. The Turtle, now feeling very exposed, straightened as if to stare down and submit Red. The Turtle began melting at the feet and she quickly fell to the ground. The melting of skin and bone continued up her legs as she screamed in agony. At the knees she was writhing and incoherent and by the time her hips were melting, she was dead.

Red willed her body to cool then walked into the house to find Tobias trying to pry Joel's hands from the nails that pierced the palms. Tears streamed down he faces of both men who were completely unable to comprehend how their own government could do this. Red was incapable of crying, but entirely capable of rage. She melted the nails with a push of a finger, cauterizing the wound, then carried Joel to his bed.

"Are you OK?" asked a frantic Tobias.

"Those who did this have been judged. Those who ordered it have yet to be sentenced."

"Red, this is not the way of this world, you will invite the entire world against you."

"I am now the Lord of this world," she said. "I see my path now. I understand why I am here."

McGill sat in his room on Edwards Air Force Base. The television was on, but he was not watching it. He was thinking about the chase. The chase began as an escape that was neither dangerous nor was it fearful. There was no indication that the Alien was doing anything other than looking for a way to survive on our planet. Her host, Mr. Burlando, had not gotten sick and there was no apparent radiation. There was nothing other than her ability to fly and create intense heat that seemed to be a threat. He knew why they ran. The Alien would certainly be locked in a cage and subjected to a life of testing and experimentation. There was no reason for them to believe otherwise. They had to run, and McGill had to chase. This was the part of his job that he hated. When you were following orders, you had to push against all the intuition and experience that told you otherwise. McGill believed that in this moment, the Alien was not a threat to a humanity. In this moment she had exhibited compassion, connection and intelligence that could benefit all. She had befriended a person known in his community as a good man. A man with a track record of good sense. He was not a man to be bought or manipulated, neither was he likely to be bullied. There was no indication that he was smitten by her or otherwise manipulated. All indications showed a man who found a woman in distress and is seeking to protect her. McGill thought that he might do the same if places were swapped. The true threat were the politicians that would attempt to use this woman's power to fuel their greed, war, and dominance.

Now they had pushed her too far. They had cornered her and attempted to capture her. The gloves were off, and she was not going to go willingly. The harder they pursued and dogged her, the more likely she would turn on them. If she needed a reason to go to war with this world, no doubt our government would give it to her.

McGill's phone rang and he saw the number was unlisted, not uncommon in his line of work. "McGill" he said into the receiver. In seconds he was out the door, beating on Troutman's adjoining room and then skipping three stairs at a time to the pavement below. He backed the Suburban swiftly in an arc, tires squealing. Troutman jumped into the passenger seat and they sped down the highway.

†††

Tobias Burlando met the Suburban far up the drive, near the entry gate. He had gone outside to get air and found himself walking three miles up the dirt road, then waited until he saw the headlights of the approaching vehicle. Without a word, he got into the backseat and the car sped down the dirt to the main house. As they came through the second gate, Tobias said, you will need to slow down. McGill did, in time to see the puddles of human flesh littering the front of the house, and the still-burning hulk of helicopter.

"What the hell," started Troutman. McGill turned in his seat to look squarely at Tobias, whose eyes were large, moist, and pleading. "Is she still here?" asked McGill.

"No sir," was Tobias' reply. A woman with a cane came with anger and hate. She hurt Joel very badly. When Red discovered it, she returned.

"Where is the security detail?" asked McGill.

Tobias simply gestured his hand towards the roasted hulks. "And the woman with the cane?"

"She got what was coming to her," said a tense Tobias.

Is there anyone else here?" asked Troutman, still dumb-founded.

"Only Joel and I," was the solemn reply.

Troutman, weapon drawn and in hand, searched the house and grounds. McGill, feeling no such threat, sat at Joel's bedside while the men told their tale, beginning to end. McGill was fascinated at the story, and although understanding of the circumstance that these two simple men found themselves in, there was a grim reality of their participating in this serious event. He told them so, and in doing so became complicit in their crime. It was not his place to make these assessments. He was simply a collector of information that was passed onto those who would use this information as they saw fit. However, he found himself at odds with this training and his course of action. McGill walked out of the room an on to the porch where Troutman was biting his fingernails and spitting them into the yard. "She's gone, but I think she will be back soon."

Troutman stood squarely in front of McGill with a sour look on his face. "What do you mean gone? Get your ass back in there and find out where it is."

McGill saw the ugly side of this man for the first time and realized that there was a bad side to who he had started to think of as friend.

McGill again began to explain that after saving her friends and dispatching the Turtle and the security detail, Red was gone.

"I heard it the first time. Tell me something new."

"Nothing new to report sir," was McGill's cold reply.

"If we lose our target, I will ruin you," said Troutman through clenched teeth. "Find her. Now!"

"Yessir," said McGill calmly, knowing the only thing an explosive personality could not stand was a calm response. McGill looked straight into Troutman's eyes until Troutman spun on his heel and walked away.

Through the entire conversation, Red watched and listened from her perch high above the Ranch. She watched as teams returned and cleaned up the bodies and the wreckage. She listened and observed as new interrogators were brought in and her friends were subjected to more questioning. They gave up nothing and the new teams left within days. Troutman and McGill resumed their residence in the old caretaker's house, Tobias and Joel were left with a much larger security detail.

†††

We live in an electronic age. Information is no longer passed along on the written page; it is sent through the air. Everyone likes to believe that their information is private, but it is not. It is open to the world. Every major power in the world had the satellite images of the Alien falling to earth. The information had been intercepted and sold on the open market within hours of her impact. Everyone was watching the Americans chasing this Alien. Their scientists were busy working on tracking solutions, they were moving assets, positioning their people, and waiting for an opportunity to step in and capture the Alien for themselves.

It was not the Russians or Chinese who made the first breakthrough, it was Australia. Entirely underrated as a spy organization, even within its own government, ASIS (Australian Secret Intelligence Service) had received the information from The Australian Geospatial Agency who had been tracking individual meteors likely to impact Earth. Red had bounced off their satellite, knocking it completely out of orbit and sending it into space, but also providing amazing close-up photographs of her. Scientists had been working around the clock to determine who she was and what she was doing out there. General consensus was that she was human, and Alien conspiracy theorists were outliers. Despite their early knowledge, there was no way to tell where this object would fall, and no one believed it could survive the descent and impact. It was assumed that it would simply burn up on reentry. As data came in from America regarding the survival of this Alien being, those same outliers became extremely popular in science circles. Since there were no apparent visual differences from the frozen form of Red to a human woman, the Australians really had no hard information to sell or barter with. They would follow along like the rest of the world to see how this would turn out.

†††

The whump-whump of choppers woke Tobias with a start. He jumped from his bed and peered through the windows of the cabin in time to see the minigun mounted to the lead chopper unleash on their security detail cutting them down. There was no sign of the sentry's, each having been dispatched by ground troops. A black helicopter was racing up pasture, sweeping into a swift landing. It was followed by several large helicopters hoisting a large net of stainless steel. To his right, a convoy of trucks had disgorged soldiers who were spreading out around the ranch house in tactical fashion. "Red!" He cried out in his mind, reaching for hers.

Red's temperament changed immediately, and he felt a flash of heat into his mind. "Red, what are you doing? Red!"

McGill and Troutman bolted from the caretaker's home and ran through the forest towards the main house, unobserved by the convoy. Their weapons were drawn and their significant experience in warfare directing their unified action. Joel stepped onto the porch of the main house carrying a shotgun, missed in the previous search of the property, as the two agents joined him, ready for a battle. Quickly surrounded, they lowered their weapons and raised their hands.

From behind the convoy, in plain view of the men on the

porch, they saw the full magnificence and glory of the Red Child as she dropped from her perch and spread a carpet of flame across the soldiers. As weapons were brought to bear, Red blasted the heat of the sun outward, completely reducing those nearby to ash, and ruthlessly blinding and scorching those that had cover. She walked among the soldiers in the meadow with light emitting from her eyes and fingers, cleaving bodies in two, removing limbs and reducing a platoon of seasoned troops to a mountain of death in mere seconds.

In the open, the trap was set, and the cable net draped over her body. The tons of weight and force of the net was expected to force her to the ground where she could be subdued. It failed in spectacular fashion. As the net draped, she simply melted it. She looked upwards and with a blast of heat energy, she disabled each of the helicopters, causing them to immediately fly off towards the valley below, smoke trailing their path.

Their plan thwarted, the lead helicopter increased throttle and attempted to take off. Red glided casually in front of it, stretching out her arms and laying out a field of fire, testing the resolve of the pilot. The aircraft descended to the ground and powered down the rotors. A man dressed in tactical black stepped out of the helicopter. Red walked close enough for the man to feel the intensity of her heat and wrath. He shielded his face and turned away until she diminished her intensity.

He smiled a cruel and arrogant smile of one with the upper hand. The man looked around at the carnage surrounding them and gave her an approving look. "You are far more than we ever expected, simply magnificent," he said.

"Who are you?" asked Red.

Perhaps we could discuss th…"

Red blasted the helicopter with a massive beam of light emitting from both palms outstretched. The helicopter rolled across the pasture, shedding its rotor and tail as it went. It ended in a crumpled heap against the opposite hill.

The man finally recognized the weight of his predicament and began to stammer out the reasons she needed his wealth and power. Red stepped towards the man cooled her hands and placed them on either side of his head. She reached inside his mind and tore it from him. She knew everything inside of him before shoving a thought into his crippled mind," I am the Lord. I have judged and found you guilty. I have executed my judgment." The man's eyes opened wide; his mouth slack-jawed until smoke poured from his ears. When she released him, his lifeless body fell to the earth, cooked from the inside out.

The four men stared at Red, dumbfounded.

Red faced the porch. They could feel the heat coming off her. It became deathly quiet. They could hear the hissing and popping of her heat as she cooled down. She was floating towards them, her body shimmering, her eyes still ruby red. As she approached the vehicles, the paint began to smoke and curl. The canvas tarps covering the truck beds smoked and charred but did not ignite. Although it was obvious she was cooling, the heat was still nearly unbearable. Troutman flinched first and moved behind a bank of equipment to shield the heat.

Red was a resplendent crimson. Her aura would put any priceless ruby to shame. She remained in her gas state as she touched down. When she walked it was as if a mirage had come to life, rippling and glistening. Tobias could feel the heat coming off in waves, and Red, feeling his discomfort quickly returned to her cool and solid physical form. He looked upon her naked form in transition and he saw her for who she was. A true child of the universe. A goddess among mortals, yet as simple as any mortal he knew. He wanted to rush to her and hold her, but he was unable to comfort a woman of fire.

As she neared McGill, she turned her heat up and his skin began to burn. "I cannot survive your heat," he said as calmly as possible, but his voice betrayed the pain he was feeling. In an instant the heat was gone and Red stopped walking. Her dark pink body was beautiful, but it was her eyes that captivated him. Kaleidoscopes of flame, in a swirling milky red.

McGill spoke first. "Who were these men?" he asked waving to the charred and cleaved bodies around them.

"The last man was a Chinese Industrialist. He hired this team to capture me." McGill could hear her clearly, although her lips never moved.

McGill looked around to see if anyone else heard her reply, but apparently, she was speaking only to him. He tried it out for himself, "Do you prefer to speak mind to mind?"

"I prefer for you and your men to leave now," was her reply. "Before you join these others."

"Why did you act with such force?" asked McGill.

"They came upon us with violence," said Red. "It was not a wise action on their part."

"I see," said McGill. "I assume you could dispatch all of us in the same way?"

Red did not reply, she turned and walked towards Troutman. He fingered the trigger on his handgun and then quickly dropped it as it got red-hot. "You will allow me to access your mind," she said. Troutman started to object and Red drained the heat from her hands, grabbed his head and extracted what she needed to know. During this encounter she turned and looked directly at McGill, as if challenging him to object. When she was finished, she told Troutman to go rest, and like a drunkard, he stumbled up to the caretaker's cabin.

Red turned towards McGill. "And now you. Will you allow me to access your mind?"

"I cannot allow that," said McGill. "My mind belongs to me. I cannot and will not share it. Was it necessary for you to take what you wanted without Troutman's consent?"

"I offered and he declined. Mr. Troutman is not a smart man. Our conversation was expedited. He does not think much of you Mr. McGill. He feels that you are too soft. That you will not do what must be done. Are you soft Mr. McGill?"

He though a moment. "I gather information, observe, and learn. I am neither Judge nor Jury. I am an investigator who reports my findings. What others do with that information is not my business."

McGill saw the face of Red soften at this comment. It broke into a smile. It was a good face, he decided.

"Mr. McGill, you care very much what happens with your information. Unlike Mr. Troutman, you have a conscience."

"And how do you know this?" asked McGill.

"Troutman told me," she replied with a smile. "Mr. McGill, I must know what you know, and you must know what I have to share. You really have no option but to allow me access to your mind. I would prefer that you offer it freely than to simply take it. If it is freely given, I will share with you. If I must take it, you will get nothing but a severe headache in return."

This put McGill into a precarious position. He had information to protect, but to understand his adversary was critical, particularly if she did not yet know others of her kind had been found. "I will allow it on three conditions," he said. "First, you may only take what you need to determine my character and honor. If you believe I can be trusted, you will share the information so I can determine if I can trust you. Finally, you must give your pledge to not take more than I offer."

Her response was reasonable, "If we get through the first two, the third will determine itself."

"Let's begin," he said.

As calm as McGill wanted to be, he still flinched when her hands touched him. Her wands were warm, but not uncomfortable. She was obviously moderating the heat in her body. He would remember that. The touch was brief, and he felt nothing. She looked at him as if to say, "ready?" McGill blinked twice slowly, then nodded his head in affirmation. She placed her hand on his forehead and McGill would never be the same.

Over the next several minutes, McGill gave her access to his entire mind. His childhood, his rough teen years, and court-ordered military service, and all of the good and bad things he had done in his life.

In return, Red shared what she had to give. When she was done, McGill shared the current operation and their intelligence on Red and the others discovered in Mexico and Siberia.

†††

Once Tobias and Joel had moved the unconscious McGill to the couch in the house, they returned outside to the vast plain of carnage. As the stood surveying what Red had wrought, a new convoy of military vehicles pulled down the lane and stopped fifty yards from the house.

The troops flowed out of the troop carriers, moving into the forest for cover and spreading out in a radius around the ranch house. Red instructed the men to go inside, then she descended the porch alone and stood quietly waiting for a sign of peace or aggression.

In a clear voice she said, "who leads this group?" A woman with clusters of bronze on her shoulders came to the front. Confident, bold, and brave. She stopped between Red and her troops and as was authorized by her authority, she demanded Red's surrender.

Red surrounded herself and the officer with a circle of flame, which began to widen, forcing back the troops. When the circle was large enough, Red stepped forward. "Do you know who I am?" asked Red. The women, fear stricken nodded yes. "Do you know what I am capable of?" Again, the women nodded yes, unable to take her eyes off the fiery kaleidoscope eyes of Red. "State your mission," said Red.

"I am Colonel Grace Washington, United States Marine Corps, Camp Pendleton," was the reply.

"Is the United States Marine Corps my enemy?" asked Red.

"Ma'am, I have orders to place you under arrest. I am told you are a danger to our people and our world. These are my orders."

Like McGill, thought Red. These people were not allowed to think for themselves but were followers. Behind a wall of flame that prevented the soldier's from seeing or acting, Red reached out and placed her hands on either side of the Colonels head.

†††

The Colonel surveyed the mass of human dead around the home, and the vehicles that were smoking or melted to the earth. The sheer destruction of those who had come to capture the Red Child was frightening. It reaffirmed all that the Red Child had shared with her. Now, they both sat silently watching the soldiers fill body bag after body bag.

They moved burned out equipment into strategic formation to establish defensive positions around the entire valley.

Red was bulging with heat energy and feeling strong. Turning to Officer Washington, she said, "You will protect this property and my friends. No one, not even your own government is allowed to stand here. This is sacred ground. Do you understand?"

"Yes, My Lord," said Colonel Washington. "Please be careful as you travel to Mexico to find your Damned. I don't know if they will trust me with additional information after this."

Joel walked down off the porch carrying Red's black suit and her bandolier of gems. "You come back safe; you hear?"

Red spoke to Tobias and Joel as she donned her suit. I must go now, but I will return shortly. McGill will protect you. If he cannot, you will call to me. Then she lifted into the air and disappeared to the South.

†††

On the orders of Colonel Washington, air defenses and troops were moved forward and were on stand-by twenty-four-seven. Ground assault teams were positioned around the perimeter of the ranch and another position twenty-miles out. Colonel Washington spent hours on the phone with Edwards, the Pentagon and Washington. She spared no words in detailing the specific instructions given to her by the Red Child, as well as her commitment to ensure the safety of both the Red Child and the occupants of the Ranch.

McGill made the call to Alpha. He described the attack by the Turtle and the inevitable result. He gave a full report on his actions and those of Troutman. He was given instructions to maintain position and wait for further instructions.

McGill and Troutman sat under the shade relentlessly working on data collected from the satellites, aircraft, and instruments on the ground. They engaged in deep discussion and strategized what the next move would be. It was obvious that they could not capture Red, she was far too powerful. They could not destroy her; she was too valuable. They would have to emotionally connect to her and encourage her to cooperate. Troutman was gung-ho to escalate an offensive, but McGill knew that she was far too smart to fall for their ruse. Only genuine concern and cooperation would work. Unfortunately, that was not the government way. Politicians who were in the loop all had opinions until they witnessed the footage, then they became eerily quiet. Politicians like an easy solution and balk at anything that could blow-back on them. The short answer was always to destroy, but Alpha would never allow that to happen, and McGill was learning that Alpha had far more power and control over government policy that he would have imagined.

†††

In the early hours of morning McGill was awakened by Colonel Washington. Quietly, she whispered that Alpha was attempting to breach the first outposts and demanded safe passage for their aircraft to land at the Ranch. They were being denied entry but within minutes, the Pentagon would override the Colonels orders. They would be on site within the hour.

The first thing that happened when Alpha arrived was

that both McGill and Troutman were ordered to return to headquarters for formal debriefing. McGill attempted to contest the order, but it was clear that Troutman was not only ready to go, he likely put this order into motion. Looking at Colonel Washington, McGill and Troutman were ushered to a helicopter and was whisked away to Edwards Air Force Base.

†††

When McGill regained consciousness, he was being strapped to a stretcher and wheeled into an ambulance. He could hear the helicopter rotor in the background. He looked to the right and saw Troutman walking toward the black suburban. To his left, Lt. Comden leaned over him to look him in the eyes. He weakly nodded his head to show he was ok, then closed them again.

He woke in clean white sheets, in a clean white room. Edwards Air Force Base was a Mojave-desert shithole, but the military spares few expenses when outfitting a hospital. Lt. Comden touched his arm and waited for him to clear the cobwebs. "Troutman?" was his first word.

"He is waiting for you to wake. You know it's standard policy to dope you up when they want to interrogate you. You did alright. You were consistent with your story. Greems arrived an hour ago and he is on pins and needles waiting for his shot at you."

"Greems?"

"The Turtles second in command."

"Get my clothes, I will be ready in 10," was his reply.

"Are you sure, you are still pretty pale," she noted with

genuine concern. His response was to swing his legs off the edge of the bed and wiggle his toes.

McGill's optimism fell a little short and he was wheeled into the situation room in a wheelchair. His legs felt like rubber bands and his head began to throb the minute he sat up. Lt. Comden rolled him into the room, then stepped back to the door like a sentry.

McGill knew the interrogation would be difficult, but he still believed that his government and country to be the best in the world. He still believed in justice and innocence until proven guilty. He had bought into all these mantras that were taught to him since childhood. He was about to learn just how brutal and ruthless his country could be. The room they had selected to debrief him in was small and efficient. It looked every bit a torture chamber. All that was missing was the hooded executioner. Unfortunately, the first person to walk in the door was a soldier wearing a balaclava. "Great," thought McGill.

Following the hooded man was Greems. Thin, balding, and rat-faced, he looked haggard and worn by the events of the past few weeks.

"Begin," Said Greems.

"I assume you know what occurred up to her dispatching of the Chinese?"

"You say *her*. Are you referring to the Alien?"

"I am," was McGill's reply. "She has the presence and features of a female and both she and her companions refer to her in the feminine."

" And you refer to the Chinese?"

"McGill knew this was going to be. Long day. He began, "based on the appearance of the attackers, the makes and models of their weapons, clothing, and equipment. It is fair to assume they were Chinese. I do not believe they were a sanctioned military unit of the Chinese Army; I would presume they were a private military outfit based on the quality of equipment they possessed. Besides, he added, the Alien read the leaders mind and got the whole story."

"Continue," said Greems.

"The Alien requested that Troutman share his mind with her. He refused and she apparently took what she wanted by force. She held her hands to the side of his head and seemingly sucked him dry. I can say with authority that she knows everything he knows.

I was her next victim and we discussed the terms of her doing the same to me and it was clear that it would either be volunteered, or she would take it by force. I offered an exchange of information and she agreed.

Greems looked horrified and close to rage. "You did what? You allowed her to access your mind?"

"I exchanged information with her," he said.

Greems did not seem to hear him. You gave up Alpha to an Alien? You betrayed us?"

Trying to keep calm, McGill looked Greems square in the eye. "Did you miss the part where she already downloaded Troutman and intended the same for me? I negotiated an exchange and I gained valuable intel," he said.

"You disappoint me," spat Greems. "You have violated every principal of military conduct. You have betrayed your fellow soldier and your country. You have collaborated with the enemy."

McGill must be careful. He responded, "You know what she is capable of. By the look in your eyes, the last few days have escalated events quite a bit. She did not come here of her own choice. She is stranded here. She is struggling to understand our ways. She has befriended a handful of good people like Tobias. These people give her hope. She has also met evil people. People like you and me. The more we pursue, the more she protects. I understand you are tasked with controlling and using her as a tool, but she is far more intelligent and far more powerful than you can imagine. Left alone, she will provide a benefit to our world. If pressed, she will remake our world in her image."

"So, you are saying she is a threat that must be managed with force?" pressed Greems.

McGill could not help but roll his eyes. "I said nothing of the sort. Don't manipulate my advice to get an answer you need. Listen to my advice for the message that is there," said McGill.

"Tell me where she comes from," demanded Greems.

"She comes from another planet. She drifted through space and crash-landed here," said McGill.

"Tell me what her capabilities are."

"That is unknown, to both us and to her. Out atmosphere is different from her own. She is still learning what she can do, but with what we know, she will not be taken without significant collateral damage."

"I do not care about collateral damage" seethed Greems.

"I am tasked with capturing and containing this threat. If possible, determining if we can weaponize her. How do I achieve this goal Mr. McGill?" he fumed, barely containing his anger.

McGill burst out laughing, utterly uncontrolled. "Mr. Greems," he offered, dripping with condescension, "if this is your goal, you have already lost."

Greems raised a puny fist as if to strike McGill, then skulked out of the room. "I want him behind bars," he shouted on his way out. "He is compromised, and he has compromised our very existence."

"Interesting" was Red's response. She had been observing McGill from afar as he was debriefed.

"He is not much of a conversationalist" thought McGill. "What now?"

"Apparently you go to jail now," she said smugly. "They will come again and get a formal debrief and you will have another opportunity to find out about the others. I must know where they are. They will be scared, and scared is dangerous."

"Two have been found. How many more do you think will make it here?" he asked.

"It is unknown. We were all blown deep into space. We could land anywhere. The two out there are in danger, and if cornered and powered, they will protect themselves. I must get to them both."

"After a brief moment of quiet, McGill offered a suggestion, "the Siberian entity will likely be cold and difficult to move. The Mexican entity is near heat and may be charged. She may be able to help you."

You assume it is a "she," she said, but your reasoning is good. I will need more power if I am to make it to Mexico. I will contact you when I return." Where will you get more power? he asked, but she was already gone.

Red had returned to San Onofre to find a battalion of military surrounding the facility and crews working to repair the damage done on her last visit. She loved the rush of heat as it came to her from this powerful station and she felt she could use a top-off for the long flight ahead. She landed in the back of a seedy hotel next to a Chevron station with a good view of the facility, the beaches, and the nearby freeway.

She walked down the rutted, uneven walkway to the weedy, trash-filled empty lot next to the hotel. She lay down flat on her belly, arms and legs outstretched and she sent a powerful bolt of energy down and outward into the earth. She was looking for any sub terrain ways to get into the facility. She got to her knees and brushed off the debris from her body. From behind she heard footsteps. She turned to see a man standing, backlit by the neon of the hotel. The man said nothing, just stood there looking at her body. He was stocky, bordering on obese. He was dirty and unshaven. On any planet he would have been threatening and ugly. He walked around to face Red, who was still on her knees. The man reached slowly with his right hand and pulled at the tail of his belt, first left to unhook the key, then right to allow it to fall. He unbuttoned the snap closure on his jeans and then rattled the zipper down. His eyes never left the body of Red.

He reached out his left hand and placed it firmly on Reds smooth red scalp. His eyes were little more than slits now as his fantasy dream knelt before him. He put his thumbs into the top of his jeans and pushed down, exposing himself to his victim. Red knew what Tobias knew about the Earth and its underbelly. She understood clearly what his man intended. She could see the clip of the pocketknife as his jeans slid over his hips. Red started to stand, but the man swept his backhand across her face. That was the moment Red looked up directly into his eyes, and he into hers. His eyes grew wide at the kaleidoscope eyes of red. She floated effortlessly to her feet, and then to eye level with the brute. Her eyes never left his, nor his hers. Standing, they were eye to eye, hers full of heat and fury, his wide with fear. He slapped open-handed at her head and the skin of his hand skin stuck to her head and began to smoke. He tried to remove it, but he could not. He tried to scream, but no sound left his throat. The flesh ran off his hand like liquid, down her forehead and over her nose and mouth, the bone charring black and then to ash, sugaring the liquefied flesh on her face. Red placed her hand on the man's chest, fingers splayed, and pushed a perfect cookie-cutter impression of her hand through his chest and out the other side. All the while, the man could not stop looking at her eyes. She left him as a pile of ash, tried, judged, sentenced, and executed.

†††

Red simplified her problem by watching where the cables left the power plant and dispersed power to the community. She did not need to go into the power plant. The power plant would come to her. She rose up and crossed over the freeway and into a copse of eucalyptus trees that were bisected by the high-tension lines. She went to the bottom of the first tower, raised her arms, and sucked the heat energy downward through the air, drawing deep of the electrical nutrient her body had depleted itself of. This was not as efficient as direct connection, but she could recharge in this manner anywhere, perhaps even in flight.

Once again, the engineers from the plant saw the power drain and they alerted the military that the Alien had returned. Red would be supercharged and long gone by the time they traced the power discharge to a location one quarter mile from the power station.

Red had no plan for her flight to Mexico. She did not understand what resources were there so she would have to move carefully. She reached out to Tobias. I need to know where I am going in Mexico.”

She envisioned Tobias at Joel's old computer, bringing up a map of North and South America. “Can you see this,” asked Tobias? “I don't want to enhance the image until I have to. I know they are watching,” he thought.

“Mr. McGill?” interceded Red so that Tobias could hear as well. Are you OK?”

“I'm fine Red. Don't you worry about me yet.”

“I need to know where the Damned was found in Mexico,” she asked.

“The Damned? What is a Damned?” McGill asked. Tobias was curious as well.

“It is the term we use for those of our planet who serve the Lord.”

“Like angels or demons,” asked McGill?

Red was losing patience and so she sent a mental image to both men that clearly defined the life, role, and place of the Damned in society.

Dear God in Heaven," said McGill. Is that where you come from? It looks like the surface of the goddamned sun."

"Mr. McGill, I am a Child of the Sun. I have served with the Damned, and I have led them. I am now the Lord of those who have found this planet. I must protect them, but also teach them the way so that they do not find the wrong path. It is entirely possible that they could imprint on the wrong human, just as I imprinted on the goodness of Tobias. It is imperative that I rescue my Damned before they are captured."

"The place is called Popocatepetl. It is an active Volcano South of Mexico City."

"How far," she asked.

"It's about 1900 miles. If it helps, it is a10 hour flight by commercial jet."

Thank you, Mr. McGill. I will come for you when I can. Tobias, give my love to Joel when he arrives." Be safe, they both thought as Red streaked across the desert sky.

†††

The sun baked the desert, reflecting heat off the sand and warming her as she flew at low altitude. As she approached the border between the United States and Mexico, the fighter jets attempted to intercept her, and she casually avoided them. Red lifted off the ground two hundred feet to get her bearings. The first missile missed her by a wide margin. The second followed closely behind and she easily dodged it. She viewed the contrail of smoke to find their origin and made a mental note. Four more missiles were inbound, and she drifted to the side as they harmlessly flew by. She could see several jets inbound from Miramar Station in San Diego. She could stay here and be distracted, or she could just.., and she was gone. Greems watched the screen from Mission Control as the Alien disappeared, leaving a comet-tail of fire and flame in her wake. "Zoom out!" he ordered, and the screen took in a larger area. "She is headed to Popocatepetl. I want all personnel at the Popocatepetl incident on alert. Notify the Mexican President that we are coming, and he better stay out of our way."

At this speed, Red could feel her energy wane. She would not be able to sustain this rate of speed the entire distance. She slowed her speed and began a gradual ascent. When the cold of the atmosphere became uncomfortable, she descended 500 feet and continued. She had a good field of vision to the horizon. She followed the west coast of Mexico and had the Sun on her right and the moon was rising in the East, giving her three strong landmarks to work with. They knew where she was going. Did it matter if she left a trail? She did not like the idea of trying to find the Damned and fight off two countries on low power. She really did not know what "empty" felt like, and she did not want to know. Ten minutes later she saw bolts of lightning lighting up the eastern sky, but she did not want to be distracted from her course. She determined that the best action was to take small recharges from many places, just as she had done in the mountains of California.

Her first stop was a power station in Hermosillo. The building was run down and filthy, but there were workers trucks in the parking lot, and she could feel the power coming off the transmission lines. She came in from the coast, flying low, startling several residents. She dropped inside the chain-link fencing and walked straight to the first transmission lines she saw. Within a few minutes, she had siphoned off a strong current and recharged her spent crystals. She returned the way she had come, hoping that her energy draw would seem a simple anomaly.

She recharged again in Culiacan and a third time in Manzanillo. She knew that the predictability of her drawing from coastal towns would raise her profile, but she felt that if she kept moving, she would stay ahead of the wolves at her door. As she was charging the last of the crystals in Manzanillo, the sniper fired, his weapon was a net-gun, used by special forces to shoot a net at its victim, wrapping them until they could be detained. It failed spectacularly. Red felt the constricting webbing and simply melted it from her body. She rose to her full height and turned slowly. A second and third web hit her, melting instantly. That was when the first real bullet slammed into her body. Her physical presence was solid, and the bullet struck and expanded, instantly triggering her gas-form self to allow the copper slug to pass through harmlessly. Red efficiently dispatched each sniper with a beam of energy. She could now feel the presence of many more figures in the shadows, apparently waiting for orders. As they retreated, the first missile struck the electric panel to her right, shooting sparks and shrapnel in all directions. In her vapor form she was unhurt, but the explosion was an intense heat, filling her with such heat energy it would almost be considered orgasmic. She radiated a beautiful crimson light now, hyper charged with the tremendous explosive heat. Unharmed, two more missiles struck. It was such a beautiful, wonderful feeling to be charged in this way. She lay her head back, spread her arms and welcomed more, but the attack stopped. Apparently, they had not expected these results. Her vapor form could not support the weight of her bandolier or suit and so she knelt and solidified the fingers of her right hand, lifting them from the ground. With a smile and a wave, she lifted into the air and sped eastward towards Mexico City.

In Southern Mexico City, Red descended into the Favela, a slum of great poverty. It was late at night and there was a lone man curled into a tight ball on the step of an abandoned storefront. Although the temperatures are searing hot in the daytime, it was quite cold at night and she could see the man shivering in the cool evening. She sat on the step next to him. His back was to her and he had not heard her approach. Gently, she sent out warmth to him. It was a comforting warmth, like a blanket, or the arm of a loved one pulling you close. His shivering stopped and his breathing became less labored. He rested calmly. As Red sat next to him, she gently touched his greasy, disheveled hair, running her fingers softly through it as if he were dreaming. She sent a thought of peace and comfort, then she drew out his knowledge of his language. Once she understood, she asked the man if he would please share his knowledge with her. The man, still sleeping, said "I have nothing else to offer, take what you will mi 'angel." She gently sought out the geography of the region and of Popocatepetl. The man was old and has vast knowledge, although much of it was obscured by cheap mescal. He was a common drunk. She delved deeper and saw his life, which was full of loss and sorrow. She gave him a piece of advice. "Sorrow can be paralyzing, or it can be the fuel of change and greatness. Everything depended on what you did with your sorrow. His fate was in his own hands." She sobered him, removed his craving and she placed within him a strength that he had as a young man. She left him sleeping on the step, wishing him health and happiness. She lifted off the ground with no other witness than a sleeping man.

She flew low at a reasonable pace, conserving every degree of warmth her body had until she saw the smoke rising from the crest of the volcano. She quickly ascended to a place directly below cloud cover, hoping that the clouds would obscure her from below. She could her the drone of airplanes above her and as she came closer to the mountain, she could see the helicopters searching around it. Although there were military at all points around the mountain, it was fairly obvious where the Damned was holed up. A large force of military was currently engaged in a gun battle with another smaller force on the Northern slope. A helicopter gunship had moved into position and was promptly shot down by a surface to air missile. Troops we trying to set up artillery downslope but using it without killing their own soldiers would be tricky. She could see a third group trying to covertly sneak in from the West. Everyone was trying to reach her Damned first.

Red sent her mind out into the volcanos slope, searching for her lost sibling. There should have been some response, but it was eerily silent. She moved into position on the side opposite the battling forces and tried again, this time sensing, rather than hearing, a weak voice in her head. She dropped in elevation, calling out. As she neared, the response became stronger. She followed the voice, unaware that she was now visible below. All aircraft forgot their soldiers and focused directly on her. From above came several large aircraft and a squadron of fighters. Their presence was a distraction from her search, but she must find her Damned. As she descended, so did the aircraft. The artillery on the ground moved to track her descent and the combat on the North slope ceased as the ground troops scrambled to move into position on the western slope. Red discovered the lifeless form of her sibling Damned in a crevasse. She sent a message of reassurance and she dropped into the deep, narrow canyon. The Damned was ashen gray and weak. It had attempted to get to the heat of the lava flow, but in its weakened state, and with the surface crawling with hunters, it determined to stay in place where it continued to weaken. Red placed one hand on the Damned's head and the other on its belly. She poured heat and energy into the body and watched as it slowly graduated to a pale pink color. The previously black eyes were now alive and colorful. They looked at each other for a long moment before Red communicated that they were in great danger, but that from this danger might come opportunity. She continued to flow heat energy into the Damned while she explained the necessity to maintain a vapor form. As Reds energy drained, she began placing a hand over the crystals in succession. They had depleted over half of the crystals when the first missile hit. It blew the crevasse open and lava rock exploded into the sky, along with Red and her sibling. Both absorbed as much energy as possible from the blast while airborne, then, according to plan, they both rose, then dropped into the center of the volcano's cone. Through smoke and ash, the dove headlong into the very heart of the volcano.

†††

"Goddammit," said Greems. "Why the hell do those damned Mexicans keep firing on her. Don't they see that only makes her stronger? They will play hell trying to get her out of that heat, and the longer she stays there, the harder it will be to get her grounded."

"Yes Sir," said the one of the Alpha officers sitting in the command room. "The good news is that she found and resurrected a second red entity. If we pull this off correctly, we could have two, or even three of them."

"What is the status on the Siberian entity?" said Greems.

"It is frozen solid and being kept in a meat locker while our forces try to get to it. The Russians have moved in re-inforcements and our teams are having difficulty maintaining their ground. If something does not happen soon, we will have to pull back."

"And the Chinese?"

"The Chinese are not the problem Sir," he replied. The Mongolians are actually the primary threat. China does not want a war with Russia, so their teams are small and tactical. The Mongolians have moved their entire military North into Russia, effectively invading the country. They are going all-in on claiming this technology for their own."

"No kidding?" said Greems. "Who else can we stir up and throw into that pile?"

"We have several Baltic states already amassing at the border, awaiting an order to invade."

"Well, give them the order then," said the Agency man.

†††

Invading a Country with the military strength of Russia is no small decision. The Russian military had initially sent in a tactical team to evacuate the captured Alien, and they were quickly outnumbered and slaughtered by the Mongolians. The Mongols currently held the town of Kyzyl and were fighting their way to the Military installation where the Alien was held. President Nitup, in his arrogant and overbearing way, ordered Kyzyl to be leveled by bombs, after which he would simply march in and take what was his. The residents of Kyzyl were insignificant in comparison to the prize that was sealed within the city. The planes were dispatched, and this small village of 100,000 souls in Siberia was obliterated. Along with the town, the good portion of the Mongol Army, and hundreds of fighters from around the world were buried in rubble. In moments, the entire area was bombed into oblivion. In the destruction, a single bomb inadvertently fell inside the walls of the of the military base, landing squarely on the kitchen where the Damned lay frozen. The intensity of the blast peeled open the building and blew its contents across two city blocks. Amid the wailing, crying and screams of the dead and dying, a crimson red body raised its head.

†††

The beautiful heat of lava is like a Jacuzzi tub. It instantly soothes and refreshes a body. It rejuvenates and heals. As Red bathed, she vaporized and felt the heat saturate every atom in her body. With her Damned in tow, she drove straight into the center of the mountain, using the lava as her highway. She felt the impact of the bombs above, creating downward pressure. "More," she thought. The barrage continued until the pressure began to come from below. With the weight of the top of the mountain blown off, the pressure from deep underground would have a path out. She felt the current strengthen. She told her Damned to turn now and race with the pressure. As the pressure increased, so did their velocity until it became an immense explosion. Their vapor bodies blew from the volcanos cone along with millions of tons of rock, ash, and lava. The aircraft above that was not burned, became disabled by the cinders that fell from the sky like a hailstorm. Within the ash cloud, neither Red nor her sibling could see anything, they just kept climbing until they were above the dust. Even with the heat of the eruption below, it was cold at this altitude and Red and her Damned, flew at break-neck speeds to the outer edge of the ash cloud and then dropped two thousand feet, racing Northward towards Mexico City. With Mexican jet fighters on their tail, they flew into the city, then landed in the favela, lost in the slums of Mexico.

Red was glowing red-hot as she landed, with her sibling by her side. She willed her body to cool and shared with her Damned how to achieve the same. Although crudely done, the Damned was able to cool sufficiently as to stop the ground under her feet from scorching. Red looked to the step where the man had previously been. She reached out to his mind and found him close by. "It is your angel, she said, I need your help, come to me." A man who was poorly dressed, but obviously made some effort to clean himself up ran into the street, mouth agape at the sight before him. "Take me where it is safe," Red asked.

The man pointed down an alleyway and the two followed him at a safe distance. Two nude, Red women do not go unnoticed in this poor district. Although there was little technology, the news spread by word of mouth like wildfire. Antonio the drunk was leading two demons into the Church, was the message. Indeed, Antonio the drunk had led Red and her Damned into a small Church. Its only luxury was a single hand-made pane of stained glass depicting the Lord. Once inside, Antonio shut the doors behind them and led them to the front of the Church where he dropped to his knees and began praying to the figure nailed to a cross. Red sent the man her appreciation and she checked on her Damned, examining it both mentally and physically. The elderly Priest came into the room to console those few who would come into his humble Church for guidance. Upon seeing the two demons, he clasped his cross necklace to his chest, and he began reciting whatever verses came into his head. Looking frantically to the right, he saw Antonio pray-ing. "Antonio, Antonio, the demon, you must run!"

Antonio looked into the eyes of the Priest begging forgiveness. "It is not a demon Father; it is an angel."

The Priest, finding his faith, held his cross necklace out in front of him. His lips curled in anger at the violation of his

Church. "Foul demon," he commanded, "leave this Holy place."

Red, having finished her assessment of her Damned, turned to confront the Priest. "We are no demons," she said with authority. "I am the Lord, and this is my Damned. We seek your sanctuary until the danger passes."

The Priest pointed a crooked finger up to the crucifix on the wall as he spat "that is my Lord. You are as Red as the depths of Hell! How dare you claim to be the Lord my God."

She calmly answered, "I never claimed to be a God." Red glided forward, toes barely touching the floor, and placed her hand squarely in the middle of the Priests forehead. The Priest tried to pull away but found himself unable. He writhed in anguish, mentally chanting his prayers to avoid hearing what Red had to say. Eventually she sent a threat to be still and the Priest stood very, very still. "Listen to my words," she said.

The Priest was clear of mind and spirit when they disconnected. He had always struggled with the teaching of the Church. It seemed so far-fetched and unbelievable, but he told himself it was a sin to not believe and so he simply went through the rituals. For over 60 years he had been teaching in this very Church, in this small Favela, never convinced his words were real, and today his belief was validated. As the connection was released, he did not bow, or prostrate himself. He did not drop to his knees. He stood tall and proud and committed his life to this Lord, whom he knew to be good, kind, and pure. This Lord was everything his own God was not. As the aircraft flew overhead and the sounds of air-raid sirens rang out from the wealthier parts of the city, the curious and the guilty streamed into the Church to check on the rumor. They were amazed to see their trusted Priest laughing and conversing with this demon, who seems not to be a demon at all.

Antonio stood up at the lectern and retold the story of his visit by the angel. He told of the return of his angel and of this great visit, and how it benefitted the favela. The Priest had asked the name of Reds companion and Red simply noted that it was one of the Damned. The Priest said that every creature should have a name. Red thought for a moment and then looked at her Damned. "A name is given by these people to each living thing of value. Here, you are no longer Damned. You are no longer an outcast. Your sins of the past have no place here. You must focus on what and who you choose to be in this place. I do not know the sins of your past. I do not know why the Creator's named you one of the Damned. You have served your sentence and today, I release you. If you will accept it, I shall call you Vulcán, for it was in a volcano I found you."

Vulcán did not understand. It had been Damned since a child, and for many thousands of years since. She did not even remember her offense. She knew only to follow the Lord. Yet, she accepted this honor and she bowed low, sending a vision of peace to those assembled.

Tears flowed from every man and woman in the Church, which was now overflowing into the streets. As dawn approached, the Priest asked, "what shall I do with my life now that I know there is no God?" And Red told him to stay and be a beacon of hope for his neighbors. "Take care of them, as he always had. Be their father, brother, and son. Care for them as his own. Then Red took five of the crystals from her Bandolier and placed them into the Priests hands. You shall sell these and use the money to feed and clothe your people."

The Priest presented Red and Vulcán with long, hooded, sackcloth robes used by his order. They donned them, sent out a message of hope to all in the Favela and they disappeared into the streets of the slum.

They had a serious problem. They had another of the Damned on the other side of the world. Red had attempted to contact the third, without success, yet she tried again and this time she felt a powerful stirring. Vulcán felt it as well and it unsettled her. Red focused on the sensation and she sent out her image, reassuring her Damned that she was safe, and here on this planet.

The response was clear and powerful. "My Lord, I will come to you."

"Tell me where you are and what is happening," said Red.

The Damned gave her his best summation of the death and destruction around her."

"Are you whole?" she asked.

"I am not whole, but I am capable, but I do not understand where I am or what is happening."

She could sense the panic in its voice. "You must find safety," she said. "Cover yourself and slip out of the area to a remote place. Practice containing your heat energy so that they cannot find you. I have another of our family with me and we shall come for you."

She disconnected, looking at Vulcán. "I am worried for this one" she said.

"I am worried as well," Vulcán replied.

†††

Red reached out to Tobias, who was overjoyed to hear from her. She gave a short account of her journey and she could hear the concern in his voice. She described Vulcán, and her childish intellect.

"You were little better when I found you," Tobias said with a sharp tongue.

These moments of levity were what Red lived for and she loved Tobias for bringing them. "Are you well, are you safe?" she asked him.

"They are always close, but they do not want me. They are waiting for your return" he said.

"How is Joel?"

"He is here with me," Tobias replied.

Red reached out to Joel and the three of them shared one mind. "I'll be damned," said Joel. "Like a Goddam conference call here. Nothin more fritenin than being in Tobias' head."

"I miss you both, she said. Be safe, be careful and I will come to you when I can."

"How do you think you will get to Siberia," asked Tobias?

"How else," said Joel, "she's gonna fly first class."

†††

Finding an honest human is difficult, but Red had been

fortunate to find a handful of them. Today she needed something far more difficult. She needed powerful person of wealth that she could control. She connected to Vulcán and shared her plan, the risks, and what to do should anything unexpected happen. If she were going to get where she needed to go, she would need transportation. Traversing half the globe may well be beyond their abilities. Her plan was quite simple. To cross this distance on their own would be to leave a trail. This would allow their adversary to anticipate their path and return. She needed to find the right person to ferry them.

Red and Vulcán walked through the Favela unmolested in their Priests frocks. They kept their heads covered and down as they walked the miles through the worst part of the city and up into the places of wealth and privilege. The Priest had given them specific instructions on finding Don Elodios' home. As they walked up the long private drive towards his estate Red saw the cameras positioned in the trees lining the drive. Eventually, they came to a van and a truck blocking a large gate. The vehicles were filled with uniformed men, but it was the man standing in the back of the truck manning a very serious mounted gun that held her attention. "Buenos Dias," said the man in the passenger seat of the van.

"Buenos Dias," replied Red. "Permitir pasar (permit is to pass)," she said.

"No puedas pasar" was the reply.

Red removed her hood and stood before him in glistening pink. "Permitir pasar," she said.

Now the man smiled a greasy grin. "No puedas pasar" was his reply.

Red stepped it up a notch by unveiling the fullness of her presence. The heat that came from her burned the robe to ash and she stood before the man in a ruby blaze. Now she smiled, "Permitir pasar," she said.

The man was dumbfounded and slack-jawed, "Que eres (what are you)?" was the reply.

"I am the Lord, come for Don Elodios," was her reply. Permitir pasar o me enfadar'e (Permit me to pass or I will become angry)."

The man stepped from the van, weapon in hand, clearly shaking. "No puedas pasar," he said, adding "por favor."

Red slowly ramped up the heat in the vehicles until the men scrambled out of them. She then reduced the trucks to a molten puddle. As the weapons became hot, the security force tossed them into the road and stood before her not understanding, but clearly afraid. "Permitir pasar," she said again.

"Por Favor sigame," (please follow me), was his beaten response.

Don Elodios was not a good man. He was a man of business, that business being drugs and prostitution. He was the wealthiest man in Mexico, and it showed in his vast palace of a home. Vulcán was confused how so many people had nothing in the Favela while this man had so much on this estate. Red responded that not all men were good. Don Elodios met them at the door, having seen their entrance on video. Being a well-connected man, he had been following the information about the Aliens for several days and was surprised and pleased that they had fallen right into his lap. He spread his arms in welcome, like an old friend. "Hola mis Amigas, bienvenidos a mi casa" he said gracefully. "It is my great pleasure to be your host tonight, for your reputation precedes you. Many countries and many corporations would pay dearly to capture you and you have walked willingly into my home. Who am I to complain at my good fortune?" His smile was greasy, his teeth far too white, his hair far to coiffed and his clothes too trendy. He was a caricature of what a drug Lord would look like.

Red ignored the fanciful greeting and walked directly up the stairs to face this man. He watched her approach with lust. He lusted her body, but he lusted the money she would bring him more. Without preamble, Red shot out her hand and clamped it on to Don Elodios' forehead. With force she rent out his disgusting, filthy life and with greater force she shoved her demands back into his head. Releasing him, she stood stone-still to see how he would process it. His face was blank. Without emotion and with the shaky voice of a beaten man he turned to his valet and said, "ready the goose, we fly immediately."

The Rolls Royce phantom was from a bygone era. It drove over the dirt roads into the jungle like floating on a cloud, coming out on the other side at a massive airfield. At one end was a large, camouflaged hanger with doors slowly opening. As they drove inside, Red saw the massive Boeing commercial jet that consumed the interior of the hanger. The steps were being pushed into place and the plane was being fueled. The pilots were always here, living on board and ready to travel anywhere on a moment's notice. Red, Vulcán and Don Elodios' entourage walked towards the plane. Red sent an instruction to Don Elodios, who turned, face ashen white to tell his assembled that he would be flying alone with his guests today. He led them up the steps and into the luxuriously appointed cabin. The steward closed the door behind them, and the pilot nervously asked what the destination would be.

†††

When the guards came to collect McGill, they put him in manacles attached to a waist chain, which were attached to his feet in such a way as to require he walk in small choppy steps. His hands were immobilized to his sides. They led him from his cell and down the long corridor to a debrief room. He was assisted into the chair, but they made no effort to remove his manacles.

McGill was an old hand at this tactic. When his interrogator came in, he would either be released, to gain his trust, or kept manacled, to show his weakness in the face of his captors. This always played out the same way. He sat in the room alone for over 30 minutes before the knob turned and in walked Troutman. He carried two cups of coffee. So, he was playing good cop, thought McGill. Next he will order the cuffs removed. As planned, Troutman motioned to the guards to remove the shackles, feigning annoyance that his good friend was being treated this way. The guards unshackled McGill, who rubbed at his aching wrists, just as he knew they expected of him. "you take it black, right?"

"Yea," said McGill, slightly sipping at the hot coffee and tasting the chemical aftertaste of the medication. Troutman took a long drink from his own, to encourage McGill to do the same. McGill would occasionally lift the cup to his mouth, but never drank, the plastic lid concealing his trick. He would have to remember not to put lids on in the future when he was conducting the interrogation.

"So, she got you too, right?" said Troutman.

"She got me too," said McGill.

"I tell ya, it was a freaky thing, having her reach into my head and run around in it like she owned it. I could see everything she could see. Memories I had long forgotten were brought up with ease and rifled through. My entire life filleted open like a fish for her to examine. I had a lot of bad stuff in there. Stuff no normal guy would do. Stuff that should be left forgotten. She knows it all and there was nothing I could do about it. That how it was for you too?"

Troutman took another long drink of his coffee and McGill faked joining him. "Pretty much," lied McGill.

"She got everything I knew about Alpha and the operation," said Troutman. "She knows everything. I need to know what information you had that I was not privy to. What did she get from you that I need to know about?"

McGill feigned a general laziness he had witnessed a hundred times in persons who had taken this drug. "She did the same damned thing to me," he said sloppily. "Just reached in and took what she wanted."

"What did she find?" asked Troutman directly.

"She found it all," he replied.

What-Did-She-Find?" demanded Troutman forcefully.

"She found me. She saw who I was. There was no way to lie to her. No way to deceive her. Once she showed me who she was I would not have deceived her anyway. She is the Lord, and who am I? I am just a man."

"What do you recommend we do?" asked Troutman, pressing for details.

"We gain her trust and use her for good," said McGill seriously.

"Not as a weapon?" asked Troutman. Will she be a willing player with us?"

At this McGill dropped his act and looked right into Troutman's soul. "If you think back, you will know that her way is good and righteous. She can neither be tamed or used. She cannot be managed, or pushed, or herded into a corral. She is the Lord, and we can either embrace her vision for peace and happiness on this rock, or she will wipe us from the face of the earth. Your choice Troutman." He held the gaze for a moment before relaxing back in the chair.

Troutman smiled at the deception of his partner. "you know Greems won't like this."

"Frankly, I don't care what Greems likes."

"But he is your commander," said Troutman, scanning McGill for any treasonous thoughts that he could report back like a good soldier.

"Troutman, you're no good, and neither are the people you work for. You have no soul, no sense of decency. You follow blindly to climb the ladder. You get off on the recognition from your superiors, not on achieving something great. There is something great happening right here, today. You and I are part of it. There is an opportunity for us to direct this moment and nurture it toward the greater good, or we try to squash it in the name of politics, power, and money. Which side are you on buddy?"

Troutman could not answer. He was struggling with the same moral question. He had seen inside of Red and he knew what was right, but he was conditioned to follow. Rather than answer his partner, he stood and walked to the door. He rapped his knuckles twice and the door opened, allowing Troutman to slip out. There was a fear in him. The fear of know the right path, but also knowing he would not choose it. In that decision would be a consequence that he was not ready to pay.

McGill called out to Red and told her he was ready to leave. She told him to be patient. There was one more task that needed to be done.

During the long flight from Mexico City to Seattle, and on to Siberia, Red was in constant contact with Tobias and Joel. They had a plan to get into Siberia, but not out. There was no doubt that once they freed the Damned, there would be a fight and a run. In this remote corner of the world there are no nuclear power plants. There is precious little reliable power. They would all need to recharge with only Red's stock of tourmaline and some local power lines. This may be suitable for short bursts, but not likely in a high-speed chase or battle.

Tobias was at his aging Apple computer searching the internet for possible recharging locations. "There is so little power in that part of the world that I fear every station will be armed and monitored. Still, if you come in hot and move quickly, you can likely reach Japan. Once there, you can find infinite nuclear power resources. Japan has eleven nuclear power plants all within a one-hour flight. This means you they will have to guess which plant you are going to next. Even if they guard them all, it's easier to run and hide in a small country with massive power and active volcanic activity."

As Tobias talked, Red followed his eyes on the map. "We head South from Kyzyl to the Darkan Thermal Power Plant in Mongolia. Should take only an hour or so. If the plant is too heavily guarded, I will see if there are thermal springs you can use. The next stop, if necessary, is Oyo Tolgi, another Mongolian power plant just an hour South of Darkan. This is where it gets complicated. You will need enough power to fly a full four hours south, over major cities in China, to the Hongyanhe Nuclear Power Plant. They will absolutely be waiting for you there. If you are not comfortable, you can sneak an hour west to the North Korean border to an active volcano called Paektu Mountain. Again, I don't see evidence of open vents, but the hot Springs are extremely hot. Perhaps there will be enough heat to get you into South Korea to a nuclear facility."

"We have no idea how Vulcán and the new Damned will store and use their energy, said Red. What happens if they are not capable of these long flights?"

"I would suggest that you follow any major power lines you see running in the same direction as your flight path. If necessary, you could always drop on a wire, recharge, and be back in flight in minutes."

"That also offers a tactical edge," said Red. They know when and where we will hit a station, but partially recharging along the way will break up the pattern of our travel and make us more unpredictable."

"Let's hope it works," Said Tobias.

†††

The Boeing jet cruised north, across the United States and

Canada. It arched over Alaska and then West and down into Russian airspace. At each border crossing, Governments had required identification. Each time, a name was dropped that was extremely high in the local government. Within minutes they were given permission and they passed the border without incident.

As they approached Russian Airspace, it was the Presidents own name they offered. Within minutes President Nitup was patched through, nervously asking the nature of Don Elodios visit. Don Elodios understood politics. He understood money. Most of all, he understood violence. Even heads of state of the most powerful countries in the world trembled at the mention of Don Elodios name. Don Elodios simply stated that he was entering Russia on business and that his business did not concern the government. Relieved, President Nitup granted him safe passage. "May I ask your destination?" asked the most powerful man in the East.

"You will protect the Alien and await my arrival. The Alien is mine," said Don Elodios.

"As you wish," was Nitup's reply, but in his heart, he knew he could not give up this prize. There would be hell to pay, but he would have his Alien, and then he would command Don Elodios and the others who he feared. No longer would he be their little toy.

†††

Kyzyl is a small town by world standards, but significant for remote Siberia. With a population of just over 100,000 people and the only airport of any decent size in the region, it was strategically important to the Russian people. It was thought that the Mongolian people, whose border was only a few miles South of this provincial town, would never dare cross the Russian border, but cross they did. Mongolia had once been a mighty empire but was now a footnote country. Power such as was embodied in this Alien could resurrect their dominance. It was an opportunity they could not pass up.

The Mongols had met the small Russian force holding the Alien and had battered it. They were within minutes of broaching the compound when Nitup's bombs dropped and leveled the city. The destruction was total. Eighty percent of the population was killed in a single pass of the Russian bombers. With instructions to leave the military compound where the Alien was imprisoned intact, hundreds of Mongol troops surrounding the compound had survived. The lone bomb that inadvertently struck inside the compound fueled and filled the Damned with immediate and massive power. She rose from the rubble and carnage in a frantic state. Her energy pulsed wildly at anything that moved, melting, or reducing to ash every insect, rat, and human she saw. As she stepped out into the light, she faced remnants the Mongol Army, stunned by the explosions that had rocked the entire city and now spellbound by the fiery woman standing before them. Before they could react, the Damned exploded with a force that cleared the buildings around her level to the ground. The Army was reduced to dust, with only a handful of charred remains scattered about. She stood trembling, looking down at the ground. Fists clenched; knees bent as if to spring into action. How long she stood motionless she could not tell, but soon after she heard a familiar voice calling to her, reaching out. She frantically sent out a primal scream of thought energy. It was not in words, but a wail of grief and fear. The voice that returned to her simply said, "I am your Lord, I am close and coming to you."

The response was clear and powerful. "My Lord," I will come to you."

"Tell me where you are and what is happening," said Red.

"I do not know where I am. I was in an eternal cold, then a wonderful, beautiful heat, then nothing. I woke to sounds and energy I have never known before. Strange beings, strange creatures. I did not know what to do."

"Are you whole," asked Red."

I am not whole, but I am capable," she replied. "What is happening?"

Red could sense the panic in her voice. "You must find safety," she said. "Cover yourself and slip out of the area to a remote place. Practice containing your heat energy so that they cannot find you. I have another of our family with me and we shall come for you."

And the Damned had run. Across the debris and bodies. Toward the tree line and the mountains, the only place that remained standing within her vision. She reached into the information Red had given her and tentatively attempted to fly. She achieved only long bounding jumps that stretched for miles.

For hours, Red and Vulcán had been connected to the Damned. They were exchanging vast amounts of information halfway across the globe, uploading to the Damned the information she would need to survive this planet. They could feel an anger and violence in this Damned that they themselves did not possess. A defective wildness. They must correct this fault before the Damned was pressed into taking defensive action.

The bomb blast had charged the Damned. She was crimson with heat energy and had nowhere to direct it. The Lord had told her to hide, be still and lay low, but she was trembling with fearsome energy that she barely contained. Such violence on this world. Such force as was unknown on the gaseous surface of the Sun. This violence felt wrong, but it also felt right. The Damned was conflicted whether to give into it or restrain itself at the orders of the Lord.

†††

Outside of the destruction area of downtown Kyzyl, the massive jet was approaching the airport; a plane that was far too large for this small municipal station that was designed for small executive jets and prop cargo planes. A full-sized passenger jet would be hard pressed to land here, but it was on approach. The Mongol soldiers watched from behind glass panels of the terminal as the massive jet descended.

Inside, two pilots were sweating heavily. Their knuckles white on the wheel. This was madness, but they could not question the orders of Don Elodios. One choice was certain death, the other held a small chance of survival. Their rear wheels touched down within feet of the end of the runway and they kicked in the rear thrusters and stood up on the brakes for all they were worth. The jet ate up the runway, the massive wheels digging a groove in the pavement, smoke rose from the rubber and the engines screamed in defiance. Every eye in the terminal was open wide, every breathe held. The plane stopped within inches of the far end of the runway. Immediately, the pilots shut the engines down and collapsed in their chairs.

The Mongol Army had immediately moved on the jet, and as the doors of the jet opened, Red and Vulcán were out and in the air, flying swiftly towards the voice of the stranded Damned. They flew over the compound where the Damned was held captive, they saw the hundreds of bodies, charred and smoking. They continued East towards the voice of their comrade. In the distance they saw the helicopters firing at the ground. A single massive helicopter slinging a large box under its belly. To their left, a pair of fighter jets was flying at speeds equal to their own towards their goal. Red sent two arrows of energy towards the jets, disabling them to the point of returning to their base, leaving only the helicopters to contend with.

They could feel the bullets as they peppered the Damned. They could feel the icy cold coming off the box as the larger helicopter descended. They could feel the Damned, screaming in confusion, crouched down low as if to make herself small. They knew the instant that it would all end. They felt the hate and anger rise like a bullet from a gun. The Damned rose and raised her arms to the sky, never looking up. In an instant, the landscape for miles just vaporized in a ball of heat energy. Such force as had not been witnessed on this planet, washing over Red and Vulcán like a warm blanket, completely filling them both to capacity for the first time since they had been exiled from the Sun. Such force as Red knew could not exist naturally on this planet.

Helicopters, trees, rocks, even the small foothills around the Damned were completely disintegrated, the ground melted into a silica glass, as smooth and shiny as a bottle. Amidst it all, the red form standing, fists clenched, fire rippling off her body and waves of heat cooking everything within miles to ash. Even with her immense power, Red could not fathom that this one Damned could emanate such tremendous heat. Worse, she could feel the anger coming off the Damned in waves. It was completely out of control.

Red dropped to the ground directly in front of the Damned, with Vulcán close behind. She bathed in this furnace of heat. She was actually charging off the heat energy of this Damned who did not seem to diminish in energy. She seemed to be increasing its power as she spent it. "STOP," demanded Red.

The Damned made no effort to power down. "I am your Lord and you will STOP!" she again said. A faint recognition crept into the eyes of the Damned. "Quieter now, Red said "you will obey." And the intensity dimmed, and the heat subsided and the Damned stood in the center of a vast flat plain where once a forest stood.

As its sight cleared, the Damned saw her Lord. It rushed towards Red and wrapped its arms around her, its gas-body seeming to tremble. "Welcome home," said Red.

†††

"Tobias, where do we go first?"

"First, we need you to disappear for a while. A two-hour

flight south-east of you are caves. If you can get there and get underground, you can hide for a while. The best part is, there are ten exits. If one is blocked, you have nine more. It is a popular tourist area of Mongolia, but it is virtually unknown in the rest of the world. I will watch the computer screen as long as you need to see the map. It is the Dayan Deerkh Caverns."

Red looked to her Damned. "Your leaps are spectacular but will not be fast enough. You must learn to fly. We will assist you for as long as possible, but our energy resources will be diminished if we try to carry you. Do you understand?"

"I understand," she said with a calmer, more assured tone. Red instructed her Damned on the use of the tourmalines, encouraging them to reach out and draw energy from them as needed. With Red on one arm and Vulcán on the other, they lifted the Damned into the air, then mentally assisted her in the finer points of flying. Within a few minutes she was flying on her own and within the hour, she was able to maintain airspeeds comparable to Red.

Using Tobias' computer screen as a guide, Red estimated their location until she spotted the Tesiin Gol Gorge in Northern Mongolia. They dropped low into the Gorge, using the canyon walls to deflect any satellite observation. They followed this canyon for a solid 150 miles before it spilled into Jugnai Lake. The winds hit them hard as they exited the canyons. High winds that swept across the plains and the canyons of Tsagaan Uul. All three Damned would have loved to ride these winds as they did back on the Sun, but they conserved their energy and stayed on mission. From

Jugnai Lake, they steered due east, flying low over desolate desert landscape, skirting the town of Talban and zig-zagging the canyons north of Tynami Lake. Tobias had continually zoomed in on the location as they neared the caverns. From the satellite images, one can see the many nomadic trails leading into the village of Sharga. As they flew fast and low, the Damned saw nomads on the trails and steered wide, hoping to not be seen. They came in South of Sharga and banked right to the Uul River. Heading North, it was easy to find the site. No roads fed the caves, but a wheel spoke of over 50 nomad trails came in from every direction. The Damned flew to the closest trails end where they found an abandoned cavern entrance. They powered all the way down and entered the cave.

†††

Greems sat aboard his jet en route to Russia, watching the satellite feed. He saw the entire drama play out from beginning to end, watching every movement, and looking for any opportunity. He had heard of Reds destructive power but was caught completely off guard by the destructive display as the Damned was released from captivity. She watched a city block disintegrate. He let out a whoop of excitement, turning the eyes of his crew, if only for a moment. Greems could see the advance of the Russian Army. Indeed, he could see the advances of several Armies toward Kyzyl, including American forces who had come up through South Korea on stealth aircraft. He was rooting for the Aliens, hoping to buy time until his own forces could arrive. Greems knew something that these other Countries did not. He had a secret weapon.

The thermal imaging of American Satellites was exceptional. Greems tracked the three as they fled Russia heading due south. His staff had plotted every power station and heat source from Siberia to China. They would need to stop and recharge. Greems would need to know where. When the thermal images crossed Tesiin Gol, they vanished.

"Where the hell are my Aliens?" shouted Greems.

"Standby," said a staff member. Computers were manned. Every head was low scanning a screen or on a sat phone, relaying information on which satellites needed to be watching a specific location.

"I want imagery of every inch of that gorge. When they come out, I better see them," demanded the Agency head. Sweat was pouring off every staff member as the tension climbed into the red zone on the plane. Greems was swearing and throwing things. The supervisory staff were yelling, and the computer operators were frantic. Occasionally, a staff member would yell "got them!" and then would have to retract their claim. For nearly an hour the map on Greems screen showed no thermal anomaly.

"Switch to chemical trace!" shouted Greems and the imagery changed to a hunt for the chemical trail that was used in California. Slow, but he just needed a sign. The high winds in the valleys had effectively dispersed the chemical trail, and the low elevation of their flight obscured the heat signatures. To the observers on the Agency airplane, the three Aliens had disappeared in the canyons of Tesiin Gol. "Where the hell are they going?" Greems asked to himself. "Where the hell are they going?" he shouted to his staff.

Greems was on the phone to the Pentagon, requesting teams in Japan and South Korea for immediate insertion into the Tisiin Gol Canyon. He would drive them out into the open where they could be followed.

††††

The caves were warm and comfortable. Red led the way, illuminating their path. Deep inside the mountain they would leave neither a chemical trail nor a heat signature. They felt heat coming from vents deep in narrow passages of the mountain and they changed to their gas forms to navigate the narrow spaces. Down they went, into nooks and seams where no human could go, as they descended, the heat became greater. The continued for nearly three miles before they found the source. Giant caverns with intense steam vents pouring heat into the mountain. The walls of the cavern were studded with gems and crystal, grown in the heat and pressure of the earth. Like children, they wove in and out of the scalding steam, soaking in the heat. They played among the stalactites and stalagmites. They explored the vents feeding the cavern and dove deeper, flying with great speed deeper into the earth's crust. Each connected to the other. Each feeling the joy of heat, and of gas and of flight. At great depth in the earth's crust, they each found what they desired. Intense volcanic heat creating high-pressure steam. To fly in this turbulent steam current was akin to a warm bath on the Sun. After the long flight both Red and Vulcán had felt the effects of having used their energy. The young Damned did not seem to notice at all. Without a reference of time, they spent three days frolicking deep in the earth before

Red called then to rise to the surface. They raced, headlong up the narrow twisting vents, racing at blinding speeds and riding the steam at it rose. Red was the first to emerge in the large cavern. The damned was second and Vulcán, a distant third. As they came together in their gas form, they intertwined and joined together, celebrating the fact that they had survived the vastness of space and their adventures here on this planet.

The three Damned talked for most of the day. There was much to say. Much to learn. This new Damned was afraid, volatile, and unpredictable. She was immensely powerful, beyond her own control, and perhaps Red's. In a solemn moment, Red gave this Damned a name, Kyzyl. It was not only the name of the village where she was found, but the word literally means 'crimson, or red' in Russian. Red impressed on Kyzyl that she was no longer a captive Damned, but free to choose her own path. She could choose to wander this planet alone, or she could join with Red and Vulcán, protecting themselves and the humans. If she chose to stay with Red, she must answer to and follow her as the Lord. There must always be only one who is ultimately responsible for all. One to lead, two to follow.

Red was truly clear in her role on this planet. They were to be caretakers and protectors of those that desired peace and harmony. They must foster well-being and promote life rather than destroying it whenever possible. There were those that would not want harmony or peace. There are those, like Don Elodios that would create chaos for profit. These must always be given an opportunity for change, but the Lord would not tolerate the unchanging heart. If there were to be judgement, it would be done by the Lord and executed at her will by the Damned, for they were now her comrades in this strange new world.

In her summation, Red warned Kyzyl that should her choices conflict with the mission of Red and Vulcán, she would be wrapped and ejected back in space, just as their Lord had done centuries ago.

†††

Kyzyl had changed in the last few days. From a timid, scared Damned, to a realization that she had power and control on this planet that she did not have on the Sun. The words Red had used felt good. Kyzyl wanted harmony and peace more than anything. It would be a new chapter in the endless lifetime of the young Sun Child. But she also recognized that deep within her was a thirst for power and control. She knew she would follow Red, but she also knew there may come a time that they would fight for dominance. Today was not that day.

And Kyzyl, with head bowed, offered her allegiance to Red as her Lord. And Vulcán joined her, bowing low. Red, accepted their offering with dignity and welcome. They were now a family.

Red looked with love at Kyzyl and Vulcán and spoke of the love of Mer and Zel. Love that transformed the fiery star of the Sun into a garden where the Children could play. Only in contrast to this bitter and hostile planet could a Child of the Sun understand the blessings of Mer and Zel.

She spoke of the great weight and responsibility of the

Lord of the Sun. How he had borne that weight with absolute commitment to Mer and Zel, and to the Children whom he loved. She spoke of the feeling of weightlessness, of flying on the currents and of an eternal life of goodness, if it were embraced.

She acknowledged that the Sun was not a perfect place. This was evidenced by the very necessity of the Lord, and the abundance of Damned that must serve him. Indeed, Red, Vulcán and Kyzyl were all Damned. All were broken and defective. Through the example of Mer and Zel, and through the action of their Lord, they had been cast out, but in this penalty, they found a second opportunity for redemption. Not only could they strive to be like their Lord, but they must also work endlessly to this goal. They must be the ones to mold and shape this blue planet. This blue planet is not the Sun, she said. But within this planet is a possibility.

"There is no place for a noble Legion here. Being wise and good is simply not enough. I could crown you each with a veil of elements from this blue planet, but it would not satisfy the task that I must give you."

She raised her arms and created a vortex of fire, its speed increasing until it was a veritable tornado, whipping up earth and gems and stones from the cavern. Kyzyl's head became adorned and encrusted with a veil of precious diamond powder. Like a silk cloth, it sparkled in brilliance. It wove itself over her brilliant red face, obscuring it while allowing a vision of her beauty. The veil grew downward, over their shoulders, now encrusted with red gemstones of ruby, spinel, rubellite, sapphire, garnet, and apatite, forming a cowl finished with a short cape of volcanic dust. The red gems embodied passion, courage, power, will and desire. The volcanic rock embodies a grounding to the earth and dissipates anger. A dangerous, but powerful blessing on Kyzyl, the Red Holy.

For Vulcán, she made the self-same diamond dust veil that covered her face and head. Her cowl was a green web of emerald, peridot, jade, vesuvianite, opal and alexandrite. The cowl draped down her back was of black opal. Green gemstones embodied balance, empathy, and compassion while the opal is a stone of foresight and healing. Vulcán was a compassionate and giving creature. These elements would heighten this character and remind her daily of her commitment as the Green Holy.

Red drew from the heat of the cavern to charge these gems with energy, tightly packed and available to increase the health of her family.

Red stepped into the fiery vortex of cavern elements and was completely obscured by the mass of earth swirling around her. As the storm subsided, the visage of Red was revealed. Her body encased in a silken suit of light. Gemstones of every color and sort found in these labyrinth canyons covered her body. Her feet painted with the dust of volcanic rock to ground her. Her hands painted with the dust of black opal as a reminder to heal and protect rather than destroy. Her head was covered in a headdress of diamond dust, draped down her back in wavy curls of white and framing her red face. The headdress was encrusted with blue gems of sapphire, tourmaline, lapis, kyanite and benitoite, which embody joy, communication and understanding. Each stone was tightly packed with heat energy, waiting for its time of need. As the room quieted and all was still, The Red Child had been reborn as The Lord.

†††

Greems was livid. Teams of special forces had stealthily inserted themselves into the Tesiin Gol gorge. They had used drones to scour every square inch of the canyon with no luck. Now, three days later, they were out of options. Satellites had been sweeping all of China and Japan for days. There was no sign of a heat signature or chemical trail anywhere. Worse, the Russians had arrived upstream the day prior, possibly with the same idea. Neither country wanted Mongolia to know they had invaded, and neither country wanted an altercation on foreign soil. The morning of the third day, a massive airdrop of Chinese solder's, in violations of the sovereignty of Mongolia, dropped on both sides of the Gorge. As Greems viewed satellite images, he could see convoys of Chinese military driving North from both Beijing and Urumqi. He could not see their stealth aircraft, but he could see the shadows of them as they passed over the desert plain. He was now confident that the Aliens had escaped out of the gorge during the last three days. It was time to pull up stakes and hunt elsewhere. Greems made the call and within 30 minutes, there were no American soldiers on the ground in Mongolia.

Russia and China would meet at the gorge. It was a tense moment when two powerful armies come together on forbidden ground. The Russian commander demanded the

Chinese Army depart and the Chinese Army obliterated the Russian teams. No more was this a covert mission. No more would the major superpowers tiptoe around. The Aliens represented the single most important weapon of the century. Whoever possessed the Aliens would control the world. China intended to be that country.

From the moment that China fired the first round, Russia was "all-in." President Nitup dispatched "all resources" to be brought into play. He was too intelligent to be distracted by the Chinese in Mongolia. It was a skirmish and he lost troops. The endgame was the Aliens. If he could acquire one, or all, the Chinese would be begging forgiveness. He wanted boots on the ground in the first location the Alien signature showed. Any country between Russia and the Aliens would be considered an enemy of the State.

The three Damned sat quietly at the cave entrance, watching the sunrise over the Sharga Valley far below. No sound but the wind. They could hear the bells on the cattle in the foothills far below. Red reached out with her mind, "McGill, how are you?"

"I will let you know when I know," responded McGill. I have a hood covering my head, I am on an airplane, and have been for some time. I am well fed and comfortable, but I have an awfully bad feeling about all of this."

"You are approaching the Japan Coast," said Red as she scanned his location. They are bringing you to us."

"Perhaps they believe I will be good bait," said McGill.

"You will be great bait. I don't know how we will avoid open conflict if they attempt to harm you."

"I appreciate the support Red," said McGill, but you know you can't come to me. No matter what they say or do, do not come to me. You need to get someplace safe and hunker down."

"We have found a place that is safe. They have not been able to find us, nor do we see any indication that they have our trail. We came in under the cover of a windstorm, leaving no trace. If we leave, we will leave a trail and they will be on us quickly. The closer we get to Beijing or Japan, the more risk we take."

"What about recharging?" asked McGill.

"We have a good source, but it is not easy to get to. We spent three days reenergizing and I think we can reach Beijing with a single jump. If not, as we get closer to Beijing, there will be plenty of electrical opportunities for us."

"What is your next move?"

We plan to move out mid-afternoon when the heat of the day is at its highest and the wind is strong. Here in the desert, it might cloak our heat and chemical signatures. They will still follow our trail as we charge from their electrical stations, but that is a much slower process. We hope to be in Beijing before they know we are headed there."

"Good luck Red."

"Good luck to you McGill.

"Tobias, Joel?" A drowsy Tobias answered but Joel was sound asleep. Not even a visit from Red could stir him.

"It is good to hear from you Red. Where are you?"

"We holed up at the cave in Mongolia. There are wonderful steam vents deep in the earth's crust. We completely charged and stayed off the radar."

"Good thinking. How are your companions?"

"Tobias, I would like you to meet Vulcán and Kyzyl. They are my Holy, my family."

"Not the Damned?"

In this place, in this time, they have committed to providing protection, compassion and goodwill to this place. No more are they Damned, for they are reborn as the Holy."

Tobias mustered all his sincerity, respect and love and sent out a message of welcome to the Damned. The gesture was returned whole heartedly by Vulcán and with reservations by Kyzyl. Tobias registered the hesitation and offered that even humans needed to learn how to trust each other. He was pleased to meet her, and he hoped that they could be friends. Kyzyl simply left the issue alone.

"What do the news services say?" asked Red.

"That is the interesting part," said Tobias. Tensions are rising all over Asia. Governments are threatening each other, but no one knows what it's all about. Mongolia seems to be mighty upset, but they are too small to matter. Japan has issued a red alert and martial law is in effect. China is massing all of its military might in Beijing and Russia is bringing in their land weapons to Vladivostok. Russia, America, and China all have their Navy's congregating in the Sea of Japan. It looks like a powder keg, ready to go off."

"With everyone looking that direction, it there another way out?" asked Red.

"I have been watching a longshot. If you feel that Kyzyl and Vulcán are up for it, it might just work. This is what I am thinking…"

Some moments later Red was reaching out to Don Elodios. "Are you still in Kyzyl?" she asked.

"Where would I go?" said the drug dealer. "My jet barely landed on this crap airport. There was a fierce battle here between Russia and the Mongols. The Mongol solders are now stacked like cordwood alongside my plane as the bull-dozers carve out a mass-grave for them. I am waiting for a small jet from Irkutsk, but the Russians have not cleared it to take off. Nitup is playing games with me. I have a game for him!"

"Is your jet fueled?"

"Of course it's fueled! How you think I will get it out of this shit place?"

"I need you in the air now. You are flying to Beijing."

†††

With the rear wheels back up the last inch of runway, the engines whined and raced to maximum before the pilots released the brakes and launched the massive plane down the pavement. It jostled and bounced over the damage caused when they arrived. The nose lifted and at the very last moment possible, the pilots gently lifted the rear wheels from the runway, feeling the wheels dip off into the soft dirt briefly before gaining altitude. They ran low, trying to gain speed to create lift. As they approached the forest ahead, they pulled back, gaining altitude slowly to prevent stalling. When they cleared the trees at the edge of the airport, the pilots began to breathe again.

Nitup had put a tracking device on Don Elodios plane, which Don Elodios staff easily found and disposed of prior to takeoff. They spent far more time correcting the commu-nications bugs and other eavesdropping devices installed by the KGB. Don Elodios felt smug knowing he had outfoxed the world leader, still, he knew Nitup would easily track to the plane by satellite. He also knew every other country was watching to see if he led them to the Aliens as well. This was a short flight and if things went sideways, it would be much shorter. His survival, and his capture of the Aliens, was dependent on his cunning and planning.

He did as was instructed, flying straight for Beijing. Every satellite with capability was tracking his plane as it crossed Russia into Mongolia. Don Elodios was focused and intense as he made several personal calls from his secure satellite phone. First was to his Russian agents who were instructed to break Nitup's kneecaps and burn down his summer home. The second call was to his agents in Beijing. He explained the equipment and containers he would need, as well as a large cargo plane on site, fueled and ready to fly. When asked about the security detail he simply said, "everyone." The third call was to his agents in Washington. The last call was to Greems.

"Congratulations on your promotion Señor Greems," said the swarthy Mexican.

"I see you are in the air once again," said Greems as he walked to his villa in Beijing. He was flanked by an entourage of agents, each armed to the teeth.

"It would seem so. I will have the Aliens shortly. They are mine. If your government wishes to have privileges to use them, you would do well to assist me rather than get in my way."

"Don Elodios, if you have been following the intelligence you will see that there are no alliances or loyalties anymore. Every country on the planet is fighting for a piece of these powerful entities. If you take them on-board, we will simply blast your plane from the sky and follow them as they escape. We will follow them everywhere until we possess them. You know the power we bring to the table. No country in the world can stand against us..."

"Excuse me Señor. Your other phone will ring in just a moment."

Greems hated when Don Elodios did this. Before the first ring he knew it would be the President. He knew his advisors would be in the room. He knew that his family was abducted, or a nuclear weapon was nearby, or some other massively manipulative trick was afoot. Don Elodios did not reason with people. He used them. Today, Greems would sit at his feet waiting for his instructions.

As his plane cruised over Mongolia, Don Elodios had confirmation that his plan was in place in Beijing. The Aliens would board the jet as he was deplaning out the back. Once they were on board, the cabin would fill with liquid nitrogen, instantly rendering them incapacitated. They would then be moved into a freezer container and loaded onto the cargo jet that would be in the air less than five minutes from start. Brilliant. Don Elodios sat back in his leather chair, hands behind his head, smiling.

"Don Elodios," came the voice.

"Yes dear," he said smiling.

"Open the door."

Don Elodios was strapped into his chair with an oxygen mask on when the rear door was pushed open by a steward, who was tied off to a seat. Red and the Holy entered in their gaseous state, slipping in through the slightly opened door while the cabin quickly decompressed. With paper and loose objects flying around, the steward pulled the door shut with all his might, in an attempt to lock it. Vulcán reached over and easily pulled the door shut so the steward could throw the lock in place. Red moved into the main cabin which was now strewn with paper from the decompression of the room. Don Elodios eyes were wide with fright at the thought of a mid-air entry. Never would he have thought of this. So cunning was this Alien!

Red stood before him, magnificent in her presence as the Lord. Her brilliant, bejeweled gown followed the curve of her breasts, her belly flat, sculpted, and tight, her long, shapely legs and her beautiful face. Such a creature was this, thought Don Elodios. And she brought him more gifts, these two tall, crimson Amazons, each easily seven-foot-tall and gilded with a fortune in jewels. Don Elodios could not imagine their value, or how they were mined. The heat from the women was charring the interior of his jet. "If you don't mind?" he asked.

Red could read his thoughts but ignored them. She took a seat opposite Don Elodios while her damned powered all the way down and moved aside staff to take the two seats next to her.

"We are changing course," said Red. "Tell them all you are off the chase and headed to Seattle." Don Elodios hesitated. Could he fake an equipment malfunction and still go to Beijing?

"No, you can't," said Red.

"No, I can't what?" asked Don Elodios.

"Don Elodios, I have been inside your head since I met you. There are no thoughts that you possess that are a secret from me. I know about Beijing, I know about Greems. I even know about Nitup. By the way, that was particularly devilish."

Don Elodios sunk in his chair, "why Seattle?"

"It completely avoids the worlds military gathered in the Sea of Japan without raising a warning," she said.

Don Elodios punched a few keys into his computer and indeed, their new course would fly North over the Sea of Okhotsk and across the Bering Sea to Washington State. They would be followed by satellite, but this would certainly throw off most of the attention from Don Elodios.

He picked up his phone and advised the Pilot to correct their course for Seattle. The plane immediately banked left. He thought about his next move. He would leave the Beijing players in place for now. He needed to throw Greems off his trail. By now he would wonder what Don Elodios was doing. It was clear in his prior call that he was headed to Beijing. This would only make him more curious.

He called Mexico City and spoke to his lieutenant. "Thomisino, you will do this for me. Lock my brother Ramon in the cellar, then leak it to the press that he is shot dead in Seattle. Yes, my brother Ramon. Do it now."

He hung up the phone. Thomisino would not fail and within a few minutes Ramon would be locked up tight. Greems will hear of his death and believe I am returning to Seattle to seek my vengeance. This is the only way, he thought.

"That was cunning," smiled Red. "I see why you are considered the master."

†††

Greems bought the act. His first call was to the President to advise him that the drug dealer was out of the picture for now. He had been monitoring the activity at the Beijing airport and had a fairly good idea of the plan Don Elodios had hatched. He moved in his own men and arranged to have a similar jet to Don Elodios' to land in place of his. Perhaps these Aliens would buy the ruse and board his plane. If so, he had everything he needed to capture and transport.

†††

Over the Pacific Ocean, Red leaned in close to Don Elodios. It was time for justice to be served, and she said so. Don Elodios was sweating. He understood exactly what these beings were and where he was in the pecking order of priorities. He knew they would not respond to money or threats. He knew that they deemed him both despicable and disposable. He thought quickly of how to save his own skin. Red read the rapid processing of thoughts in his mind before saying, "You have only one choice."

He cowered and cringed at the thought that she had the power to enter his mind. If he had this power, he could rule the world. Damn her for making this so difficult!

Vulcán and Kyzyl watched as Red placed a warm hand on top of his wrinkled and cigar-stained fingers. "You have made a life on the pain and suffering of others, she said. You have enjoyed it because you are defective. You feel powerful when you reduce others to helplessness. Today, you shall learn what it feels like to find joy in the happiness of others. You will be given an opportunity to embrace this feeling or reject it. Our judgement will rest on your choice."

Don Elodios did not like this at all. His success, power and control all came from the most negative parts of his personality. "How can you ask a tiger to change its stripes?" he asked her.

"How can you expect a benevolent Lord to stand by while a tyrant crushes the less fortunate under his foot?" she replied.

"Where is the compromise?" he asked, looking for a loophole.

"Righteousness knows no compromise," said Red. "You make every effort to be good, or you fail. There is no middle ground."

"And if I cannot make this change?" asked the drug kingpin.

"Then we will show you how." And Vulcán and Kyzyl emanated light from their bodies, illuminating the gemstone cowls and baring their very souls. The aura of this glow seemed to hover just centimeters from their body. It was the most beautiful light, and within it was the essence of goodness and charity. To look inside the Holy was to see a purity and peace. It washed away the stress and white noise of the world until the only thing those bearing witness could see was contentment. It was the kind of peace every soul aspires to, but few ever glimpse for even seconds in life. It was an example of what could be if we let it. The lesson that Don Elodios was learning was that this peace was so easy to achieve. So much easier than all the hard work and scheming he put into creating chaos, violence, and swimming against the current. He saw himself stop midstream. He felt the water pushing against his entire body, then he simply lifted his feet and allowed the current to carry him downstream. Don Elodios had grown up in the Favela. He had fought tooth and nail for everything he had. In this moment, he felt for the first time the comfort that comes with letting go and allowing life to unfold as it will. He felt the elation of riding the current into the unknown instead of the relentless planning and forcing of life to fit into the goal you had set for it. Don Elodios was known for his cruel, forced smile. The one he offered up while sliding the knife through the ribs and into the heart. Today, for the first time since his childhood, a smile appeared on his lips that was borne of joy rather than hatred. It was infectious and it melted his own heart.

Despite this overwhelming moment of beauty, Don Elodios would not choose it. With this great insight, you would think that he would turn over a new leaf, that he would convert and use all his resources for good. But the plain truth is that human nature is inherently evil. We are not borne out of good but are the product of millions of years of evolution that has demanded we kick, scream, and claw our way to the top of the food chain. Don Elodios loved the vision of the Holy the way a man might enjoy a great meal. Once finished, he returns to his life. As they let the vision fall, Don Elodios' persistent greedy, violent self-returned. With the placid smile still on his face he simply said to her, "see, I knew you could not do it."

Red's justice was swift and certain. She grabbed either side of his head with force and sucked out the badness like a talon removes a liver. Don Elodios mouth was agape, and his eyes bulged, yet he could not scream. With greater force she slammed a new reality back into his skull. The agony was unbearable. It felt as if Red had peeled his brain from its cavity and then shoved it back in. Don Elodios whimpered and then collapsed in his chair. Both Vulcán and Kyzyl bowed their heads in reverence to the judgment of the Lord.

As Don Elodios jet approached Seattle, Red and the Holy left the airplane the same way they got on. They slipped through a cracked door and drifted down using minimal energy, heading East to Mount St Helens, while Don Elodios plane landed at Seattle Airport. Without trail or heat, the three descended, isolating the greatest heat vents from the giant volcano. In a gas form, they dropped into a vent without a sound.

As Federal agents boarded his plane, Don Elodios was found unconscious and transported to the hospital. His staff and stewards were well trained to not speak of anything that happened on his aircraft and said nothing. The plane was searched, and nothing was found. It was seemingly a clean operation.

†††

Weeks later, Don Elodios would depart for Mexico City and the Favela. For many months he would transform his criminal empire into a benevolent enterprise to benefit the poor and struggling. As his competitors saw his loosening hold on the drug trade, they moved in. There was all-out war in the streets as cartels fought for dominance. Don Elodios choked down his desire to hold onto what was his. He tried to walk the path, but a single moment of weakness ignited all the hate, furor and violence that had been pent up and contained since he met Red. In an explosion of terror, Don Elodios unleashed his rage on those that would piss on his house. He not only reclaimed what he had, he relentlessly drove deeper, assassinating politicians, executing those that stood in his way and gripping his worldwide enterprise in a cocoon of fear. Don Elodios would meet Red again, many years later. She would look into his eyes and see the insanity that had festered and welled up inside of him. In the end, she would put him down like a mad dog, only to have the pack rush in to tear apart his organization and start again. Such is the path of humanity.

†††

The Damned were content at Mt. St Helens. It was a fully active volcano and the red, viscous lava was only a short way under the caldera. They bathed in the heat and rested. They shared information and ideas. Each became to know the other seamlessly. This exposed the fractures in Kyzyl, and both Red and Vulcán worked with her to heal those breaks. The greatest help came from Tobias, who opened his mind to Kyzyl and Vulcán so that they could see the same beauty in life and humanity that had touched Red. To be present in Tobias' mind was to sleep in the bosom of a happy mother. It was peace, pleasantness, and happiness. Tobias embodied all that a good man should be. Across the miles, as he slept, both

Vulcán and Kyzyl would study the mind of this simple, beautiful man.

There was one piece of unresolved duty. McGill had been the bait in Beijing. He was shackled inside a jet on the Beijing tarmac, waiting for Aliens that would never come. Greems had played his last card, and as the days followed, McGill had starved, weakened, and eventually passed out shackled to a seat. All along, Greems felt certain that McGill's plight would attract the Aliens into his trap. Eventually, Alpha pulled the plug on the operation and McGill was discovered unconscious, sitting in his own defecation and gravely ill. McGill would be moved to an air base in Japan to recover, and eventually transported to the Alpha hospital in Oregon. On a cold morning five weeks after his capture, McGill disappeared under the noses of his guards. No trace of entry or exit. Nothing caught on video in or outside of the building. It was a great mystery.

†††

Without a sighting for months, the powers of the world did not deescalate. Tension climbed and the situation escalated out of control. War is simply the madness, ego, and power obsession by a few men, who use their power to build massive machines of destruction and then send tens of thousands of innocents to their death. In all of history, there has never been a war that was borne out of true concern or compassion for the people. Every instance is one rooted in greed and insecurity.

This is not to say that the men of power are solely to blame. The media needs a story and in exchange for one, they sell their souls. They create friction from fantasy. They stoke the coals to make a mighty wildfire out of mere embers. They cloak themselves in truth while spreading falsehoods in the name of notoriety and profit.

Lest we forget, there are the minions and foolish who allow themselves to be subjugated and subservient to the will of another. It is one thing to follow instruction at your place of work or in your daily life. It is quite another to pick up a weapon, aim it at another human being and erase them from this life because your master instructed it. Lapdogs and submissive's. They yearn for a ribbon or a medal, a token of their loyalty to an act that they would never allow in normal life. An act that they would not allow an enemy country to inflict on them.

And so, armies square off. Each self-righteous in their cause. Each following orders that reached up to the ego of a single man and his inner circle of those who would profit from bloodshed. And these armies clash, and bleed and suffer and die. All in the name of God and Country.

The first act of war was the Mongolians attacking the Russians on their own soil. It was for want of a life, long passed. A revolution that would raise the Mongol people to a place of dominance on the world stage. Russia is a proud country whose presidents have kept the fires of loyalty and penalty burning for centuries. Second was the Chinese attacking the Russians in Mongolia. A world power openly slapping the face of another world power.

When these acts of hostility met in the Sea of Japan, the focus was on the Aliens. As time passed, the focus became opaque and as each country looked over the bows of their ships at the others, they saw their own shame and incompetence in their failure to bring three girls to heel.

As they drilled on their decks and as their warplanes circled overhead, the tide shifted from a singular mission to an entire breakdown of roles and strategy. Twenty sovereign countries were packed together in a small sea, waiting to see who would make the first move.

As usual, a super-power used a minor country as their bait. Feeding misinformation from one country to another, the major player encouraged a feud between the smaller countries. Always one to take the bait, North Korea was determined to protect 'land and liberty' by firing missiles from their coast at the smaller countries forces they thought they could dominate. The retaliation was swift and complete. A swarm of missiles from many countries headed to North Korea, devastating their infrastructure, and killing thousands. Nuclear facilities were targeted which achieved the goal of the Superpowers. As the bunkers and laboratories and missile silos were obliterated, a cloud of radioactive dust rose from the blasts, carried on the wind directly towards their own boats in the Sea of Japan. Irony.

With the pawn destroyed, weapons quickly moved to target other major countries, waiting for the opportunity to open fire. This is perhaps why they failed to notice missile silos within North Korea opening. Apparently, not all the nuclear infrastructure of North Korea had been obliterated.

Two missiles were launched from within this small brazen country nestled between China and Japan. Nuclear missiles fly straight up into the atmosphere, then arc downward, exploding above the earth's surface to decimate through the explosive compression of the blast on a wide area. Every Country on earth watched the upward trajectory of the missiles. Every boat in the Sea of Japan braced for their end. Every resident of every country that was watching this unfold on CNN watched with morbid curiosity and fear for what this meant.

The drama ended as both China and Japan released their countermeasures and destroyed the North Korean Missiles at their apex, before they were armed. Both countries could have been severely damaged by the blasts and the nuclear fallout of twin nuclear explosions. This was all occurring in their backyard and it was in their best interest to work together to eliminate the threat.

In a rare show of union, the governments of China and Japan, bitter enemies for generations, allied in their resolve to remove the conflict from their shared Sea. All their combined land and sea weapons targeted the flotilla of ships in the Sea of Japan. Communiques were issued to leave and return to their country of origin or be eliminated by force.

Cowed, the ships began to turn and steam out of the Sea of Japan. One collision, one firing mishap, but no escalation as each country left the confines of this small area and parked their ocean forces hundreds of miles offshore. Not gone, but certainly not crammed together in a small, confined space.

The winner of this conflict would be China, whose ground forces swept into North Korea, capturing the dictator state in a day, with minimal casualty. Without this small country as a buffer, the countries of South Korea and Japan both fortified their borders, turning a complete and successful escalation into a new and equally threatening escalation.

†††

Whether by one of Don Elodios aircraft staff, or through satellite observation, Red and her Holy were discovered by Russia, and through spies and traitors, became known to China. Both countries sent in small teams of covert operatives to confirm the existence of their presence. Red could feel the presence of humans within their local area and she sent her Holy to investigate. Vulcán exited the cave entrance and moved west, surprising the Chinese who were live-streaming Beijing. The transformation by means of the cowl made the Alien altogether more otherworldly. Vulcán spiked the electronics with electrical energy, disabling them. She then marched the operatives to the cavern entrance and inside.

Kyzyl had met the Russian operatives in much the same way, but her reaction was entirely different. She powered up, illuminating her body, and casting red pinpoints of light across the surrounding landscape from the gems on her cowl. She lay a ring of fire around the Russians, who abandoned their electronics and weapons, hands in the air. The ring closed, herding the men towards her until the heat was charring them both ahead and behind. They were in a panic and Kyzyl enjoyed their frantic fear. She dropped her power, reducing the flames and using it to herd the Russian operatives into the cavern entrance.

When Kyzyl arrived, Vulcán had her hands on the head of

the Chinese team leader, gaining access to his language, orders, and intentions. She relayed this information to her Lord, who stood facing the Chinese soldiers as they pissed their pants and shook with fear. The Lord thanked her Holy for her service and Vulcán bowed low and stepped aside, leaving the Chinese to stand before the Lord for judgment.

Kyzyl, taking a cue from her sister, searched the Russians for the leader. Finding none she put a hand on the forehead of the closest. His skin burned at her touch and his discomfort was clear as she ripped information from him. Tossing him aside, she moved to the man on the far left. This was the leader. He flinched as she thrust her hand over his head and rent the information she needed from his skull. As he dropped to the floor, his skull smoked and the handprint was charred black, split open like a sausage cooked too long.

Kyzyl stepped towards her Lord and offered the information she extracted. Red stood staring at Kyzyl for long moments. She then rebuked her and ordered her Holy to leave the cavern. Confused and hurt, Kyzyl bowed, although not low, and left the room flying.

The Lord revived the fallen Russian and sent a comforting message of apology, which he accepted. She looked at the assembled and sent them each a vision. It was a vision of promise, hope and possibility. It was also a vision of penalty and consequence. The message was clear. Each must choose their path. And each chose their Lord. Bowing low at the waist and waiting for a signal to rise, the Lord placed upon the head of each a translucent net of iron and cobalt and named them Legion. She bade them safe travels home, where they would share her message and instructions to the governments who sent them. Afterward, they would travel the land sharing a message of hope, for the Lord had come. The men filed out of the cavern and descended the slope of Mount St. Helens, returning to their vehicles, their aircraft, and their homes.

The Lord called to her Holy and they came. Red stood before them silent until Kyzyl spoke, unable to hold her voice. "I have fulfilled the duty of the Holy. I have brought you the information you requested and presented the Damned for you to judge," said the Crimson Holy.

"Your very words betray you. For you called them Damned before they were judged. They were not Damned then, and they are not Damned now. They are humans, subservient to those who would use them as tools of evil no more. They are Legion and will take our message forward to their own communities. They have seen the Lord and they will follow."

Kyzyl was mortified. "Humans cannot be Legion," she said. "They are below the Children of the Sun. They will never be our equal."

The Lord quieted Kyzyl with a blanket of compassion and love. "We are here to serve the humans, not the other way around. We must teach and lead them. We must show them what it is to be a Child of the Sun, and through us, they will aspire to our example.

"I do not understand, but I will obey," said a contrite Kyzyl, bowing low.

"Kyzyl," said a smiling Lord.

"Yes Lord," said Kyzyl as she turned and bowed again.

"Please be gentle with these humans. Their bodies are fragile."

"Yes Lord," said Kyzyl.

"My Holy," Red continued. We have a decision to make. We are found and more will come. We can choose to stay here and face them. We can choose to find a new home, or we can come out into the open and claim our place within the society of creatures on this planet. If we choose the first or last of these, we will incur great hardship and will not likely find rest as we have enjoyed these many months.

Vulcán spoke for the Holy. "We will follow."

With the support of her Holy, Red glided to the cavern entrance and leapt off the mountain and into the world.

HOME

Joel Kerry bought a new Dodge pickup and a horse trailer. They loaded up Jenga and Tobias Burlando's few possessions in Kernville and hauled them up the caretaker's house on Joel's ranch where Tobias now called home. Each morning, they shared coffee and breakfast, then headed out to mine. They opened many large pockets of gemstone, selling a few but hoarding the majority.

Months following the exodus from Mongolia, a man walked over the ridge, cutting cross country over Joel's ranch land. Joel fetched his shotgun and Tobias rocked in his chair on the deck, smiling and smoking a pipe, a habit he had picked up to remind him of the heat and smoke of Red. Joel started getting worked up, pacing back and forth as the man crossed his pasture, walking straight towards Joel's cabin.

"You stop right there partner. This here is private property and you ain't welcome," snorted the grizzled old man.

The walking man continued forward.

"I ain't warnin you again," said Joel, raising the shotgun to his shoulder.

"Put that damned thing down before you hurt a friend," laughed Tobias. The man continued forward.

"I don't know him, he ain't welcome," as Joel pulled back the hammers on the twin barrels of the gun.

"You old fool, that's McGill, come home to help us out," snarked Tobias.

Joel dropped the barrels of the gun and released he hammers. "How the hell was I s'posed to know it was him?"

Tobias raised to his feet and walked the two porch steps down to the gravel, meeting McGill and Joel at the foot of the stairs. Tobias shot a hand out, which McGill grasped firmly. McGill turned to Joel and offered his hand, which was accepted with a toothy smile.

†††

Red, Vulcán and Kyzyl arrived at the ranch by late morning. With a few minutes notice, the men were outside and waiting when the three arrived. The men were stunned by the presence of the three. With their new bejeweled appearance, they looked even more exotic and beautiful. They marveled at the height of Vulcán and Kyzyl. They had seen visions of the Sun home world and knew that Red was smaller than the others, but to see these beautiful tall beings covered in jewels was an amazing sight.

As they touched down, Tobias rushed into Reds arms and held her tight, barely giving her time to power down. He had grown a mustache and beard, his breakfast clinging to the hair, and she laughed as she studied his face. In turn, each of the men came forward to hug on Red but kept a respectful distance from the Holy.

With greeting done, Red stood alongside her Holy, introducing them each, and then the men to them. The Holy bowed low, but Tobias stepped forward and extended his hand, palm up as a gesture. Vulcán draped her hand over his and felt warmth, comfort, and happiness. He shared a welcome to the blue planet and hoped that she would find home here. Vulcán offered a meek 'thank you' and released his hand.

Tobias extended his hand to Kyzyl, who refused to accept it. In a moment of absolute submission, Tobias bowed his head, offering her direct access to all of him. Kyzyl was uncomfortable and it showed. She did not want to like these humans. She did not want to feel safe with them. She had woken to their assault. She had seen their death. She did not want any part of it, yet here was a human offering to bear everything he was to her.

With Red nervously watching, not knowing how this would play out, Kyzyl reached out her hands and placed them gently on Tobias' head. She saw his birth, the love of his mother and father. She saw him learn to eat, speak, and read. She felt his first kiss and his first union of intimacy. She saw the pain of his father's death, the suffering of losing a wife and a child. The sadness of the loss of his mother. She saw him perceiver and continue life. She saw him save the Lord and care for her. She saw that inside him he would sacrifice all for her well-being. He did not know her, yet he would lay down his life to save her. All of this was glimpsed through a gentle touch. In return, Kyzyl sent a message of thanks, and of genuine devotion to this human who was her savior.

As Kyzyl raised her eyes to Red, they shared a moment of connection. Kyzyl now understood what devotion meant. She would give all to save this human, just as she would give all to save her Lord. These men with Tobias were his friends. He trusted them, and so, they had her devotion and trust as well. Kyzyl bowed low to Joel and McGill and they returned the bow. The circle was complete.

†††

On the morning that Red decided to out herself and her Holy to the world, she notified Tobias, Joel, and McGill of her decision. This would undoubtedly open old wounds and cause no end of trouble for the Ranch. The world powers knew that these men were a weak spot for Red and they could be used as human leverage. Red conferenced Colonel Washington into the conversation, asking about reinforcements for the Ranch.

"Hell," said Joel, "We just got rid of the last of them and you want them to come back?"

"What would you have me share with my superiors about your plans?" asked Colonel Washington.

"Tell them that the Chinese and Russians are already aware and that it is our intention to be made public, worldwide," said Red. No doubt they will see the ranch as a strategic outpost. I doubt they will attempt to storm it again. If your government is the one watching over the ranch, they have a friendly connection with us. We have a mutual interest. If they fail to protect the ranch, or in any way harm its occupants, they will be judged."

†††

A convoy of trucks and equipment arrived at the Ranch that very night. By morning, large pads were cut into the earth at each end of the valley. Navy Seabees had temporary barracks built on the pads by days end and by nightfall, troops were actively living on the property, without interfering with the day-to-day operation and comfort of the ranch. Over the next few days, outposts were built along the perimeter of the ranch and soldiers with high-tech monitors were stationed at each.

At the entrance to the ranch property, Camp Scherman, a Girl Scout camp was commandeered as a military facility and the resident of Morris Ranch Road were moved for their own safety.

Helicopters brought in sophisticated electronics and set up a station at Top-Of-The-World, a mountaintop reached by road from the ranch. As the highest point in the area, it provided clear line of sight into Palm Springs and the Mojave Desert to the North, and South to San Diego, a secure overlook for the entire mountain valley.

Colonel Washington was ushered into the ranch house where she was greeted by Red and introduced to the Holy. She detailed the changes to their surrounding property and argued with Joel over the necessity. Joel's 'goddamn ranch' had become a 'goddamn military base.' She assured Joel that his privacy and safety were paramount, and she would attempt to accomplish her mission with the least distraction possible.

"You just make sure your people leave my goddamned mines alone," he spat.

"You have my word," she replied with a wink and a smile.

"Goddamned woman got me wrapped round her finger," thought Joel. He was flustered and everyone present knew it.

On the third day, the heavy aircraft arrived. The first to step out onto the pasture were Pentagon officials. Next was the President himself. His toupee' flapped in the rotor wash and he tried to hold it on while running for the house. Last were his daughters and sons, who had come along for the ride.

Red was standing on the porch with her Holy, claiming the high ground. As the excited contingent came forward, gawking and pointing at the 'Aliens, Red could feel Kyzyl's intensity flair. She sent Kyzyl a calming reassurance and stood impassive as the group rushed the porch. Red powered up, claiming space, and driving back the hoard, to smiles and laughter of Tobias, Joel, and McGill. A General ran into a daughter who bumped into the President; it was a laughable moment. The president shot an angry glance over his shoulder. He was the god-damned President. This was his god-damned spotlight. With cameras recording the moment, the President of the United States began a blustering, loud commencement. The Red Child raised her hand, stopping the buffoon cold.

As one, the Holy stepped forward, down the steps and parted the gawking onlookers, driving them back and leaving only Red and the President. The Secret Service were talking into their sleeves, a perimeter of snipers readied, and the moment became very tense.

"Would you join me inside?" offered Red wordlessly to the President.

"I have a press corps here, and people I want you to meet," said the leader of the free world.

"They will wait," she offered, and she turned and walked inside.

The President turned, waved to the assembled group, and nervously followed Red inside.

Joel dropped a cup of coffee in front of the man whom he had little respect or need for. It sloshed over the sides of the cup and made a puddle on the table. Tobias, who had no use for the man either, but was infinitely more gracious, stepped forward, lifted the cup, and wiped the table with his shirt-sleeve, then placed the cup back down.

The President, not knowing what else to do, lifted the cup and swallowed what was perhaps the worst coffee he had tasted in his life. No one offered him the OHO honey.

Red seated herself and looked at the President. There is no place here for a press corps, she said. There is no place for your minions and certainly not for your family. This is a serious situation and I need you to start acting like a leader rather than a tour-guide."

No one was going to talk to him like that. He had made millions of dollars. He had stolen an election to lead the most powerful nation in the world and he had the devotion and confidence of the entire world. He started to speak and Red cut him off.

"You were given millions by your father. You have not grown it; you have leveraged it until there is little real value.

You did not win an election, you paid off members of the electoral college to abandon their fiduciary duty to their constituents and through this cheat, you stole the Presidency. The United States was not even in the top three most powerful countries in the world anymore, and you are not revered, but rebuked, reviled, and dismissed by every thinking person."

The President was red faced and angry. He stood abruptly, spilling coffee down his front, scalding his crotch and leaving a wet stain from his belt to his shoes. He shook a finger in Red's face and was immediately flanked by the presence of the Holy. Their heat singed the silk of his jacket and burned the backs of his legs. He stood stonily still, not speaking or moving. Secret service stormed the room and they were driven back by heat.

"Shall we start again Mr. President," said the Lord?

When Red had finished her discussion with the President, they stepped outside and the cameras were popping off, capturing the stain on the President's trousers and his frightened, cowed look. He had the look of a man whose easy life was about to become much more difficult. The President and his children were ushered off to their waiting helicopter and, flanked by two others filled with secret service, they lifted off as one and returned to Edwards Air Force Base. The Press Corps decided to stay and see what else would happen.

Colonel Washington had set up a large conference table under the giant oak in the front yard. It was littered with foods and drinks. The Pentagon men and women were seated and Red stood at the end of the table. The Press Corps were cordoned off within hearing distance, but out of the way.

"Welcome," Red offered. "The world is about to change. It will not be an invasion by Aliens, or a mind-trick. Humans will not be enslaved or attacked and harvested for food." She waited for any sign that her guests found humor in her words, they did not. The press was writing and recording every word. This would no doubt go bad. So much for Tobias' recommendation that she begin with humor.

"Each of you has something to show me, but I will only take it if it is offered. Nothing I will share changes any part of you. I do not read your mind; I do not change your heart or your conciseness. I simply share and you determine the value of my message. I will show you everything, and you will each determine the value my message has to you and those you love. Colonel Washington will ensure your safety and comfort. When you have made your decision, I welcome you to come to me."

Without further comment, the Holy stepped forward and began to glow. Their energy created a dazzling display of light and color as the jewels began to radiate. Those assembled shut their eyes and sat quietly. They allowed the vision to permeate them. Video cameras captured every second, which would be broadcast and streamed worldwide. The vision started with Mer and Zel, and the love they shared. It was the story of the Children of the Sun. How they lived, how they interacted and how they managed their society and culture. They saw the Lord in his full benevolence, and they witnessed first-hand the Legion and the Holy. They saw the truth of their world, and the saw the possibility of what could be. Inside of this was the conflict that would come. The greedy, self-serving of the world would resist. Some of them were seated at this very table. It was not the Lord's task to change a heart, only to show it the way to happiness. If would not embrace this happiness, it was her job to ensure happiness for those who wanted it. No person would be allowed to undermine the will of the people. She wished to each of those assembled the courage, strength and will to support this vision. She warned those that would fight it that they would be fighting her.

During the meeting, Colonel Washington had a large medical tent erected in the pasture. As the Pentagon officials passed out, one by one, they were gently lifted on to stretchers and lay to rest in the tent.

"How do you think it went?" asked Tobias.

"I did not take from any of them. Their knowledge and motivation's mean nothing now. They will embrace or reject what is offered. Once done, we will determine a course of action."

"Can you take on an entire nation," he asked her.

"We will take on the entire world," was her answer.

The first to wake was the Marine Corps Liaison. He rose and stretched. He felt better today than any time in his memory. He had joined the Marine Corps to make this world a safer, better place. As he rose through the ranks, he realized that he was a bit player in a game played by powerful men. His job was to send innocent boys and girls to a life they would never understand. His entire life had become a con.

Within the words of the Lord, he could see an opportunity for revolution. Not to overthrow a government, but to provide the people he loved with an opportunity for true peace.

He stood on shaky legs and marched out of the tent, accepting a glass of orange juice from a soldier, then stepped off the paces to the main house.

One by one every man and woman woke and made the walk. As they stood before the Lord, they offered to share what was inside of them. They wanted to let go of the secrets and cloaks that kept them up at night. They needed to have the Lord absolve them. Of course, this was not her task. She could not absolve, only accept, and ask them to change for the benefit of humankind.

Three of their number stepped forward and asked to see more of the Lord's vision. What was her strategy? Were more Aliens coming? Was she trying to overthrow the government or take over the world? Red has already answered these things, but she knew that they were fishing for more. They wanted a disclosure they could take back and trade for a ribbon or badge. For a promotion or recognition. The did not get it, and that was OK. These were ushered to their waiting helicopter and would be returned to base.

When the assembled Pentagon officials had all been vetted, the Lord addressed them as one. We now have a duty. Our duty is to share this vision with the world. To let them know that help is on the way. They must be reassured that their lives matter and that their government will take care of them rather than simply manipulate and use them as tools.

The Lord lifted the ground around the pasture. With a raising of arms, the ground swelled, and the Iron and Cobalt found in trace amounts was brought to the surface where it was spun into cowls, draping over those in attendance. With this gift of Iron and Cobalt, I name you Legion. You shall be the messengers of life. Go now with our blessing. Today your real life begins.

†††

The President was furious. "I don't give a shit how many troops are there. I want that property leveled, now!"

"Mr. President, there are Pentagon officials, enlisted servicemen and three Aliens that could pave our way to world domination. To throw all that away is ludicrous, said the blonde aide."

"You get your goddamned airplanes in the air and you level that place now, do you understand me?"

"There is the issue of the Press Corps sir. They are waiting for approval to print. If you do not allow it you will need to give them something bigger," she stammered.

"Bigger than flipping Aliens? Goddamn press," the President shouted. "Tell them that we are declaring war on Iran. Have our boys fire some tomahawks into Teheran. That should satisfy them until the Alien problem is eliminated and no longer a story."

"You want me to order bombs to Teheran?" The aide asked.

"Do I stutter? Get it done!"

Yessir," uttered the blonde aide.

"The aide called her husband, a vocal opponent of the

administration. She was frantic and crying, something she never did. She explained what was happening and asked her husband to leak it to Congress. She had played this game and rode this horse as far as she could. If she jumped off now, she might still be able to parlay her work into an elected office.

The husband made a single call, then sat back, waiting for the shitstorm that was sure to come. It arrived in the form of a CNN news report. Reporters were standing in front of the White House reporting on the rumor of Aliens living among us. Not a word was said about the Presidents melt-down. Within hours, the news on all networks was all Aliens, all the time.

Twenty-seven Pentagon officials returned from the California Desert and were immediately arrested and sentenced to Guantanamo prison in Cuba. In this place, they would be outside of the jurisdiction of the United States Courts. As they were processed, attempts were made to strip the cowls from their heads, but they were woven into the very skulls of the adherents. Instead of being treated as service men and women who had served their country, they were beaten, scoured, and scrubbed and thrown into concrete cells. Prisoners of the very system they had themselves created. Red allowed all of this to happen so these Legion would understand the evil nature of their own government. Once the experience was complete, Colonel Washington issued orders from the Lord to release them. There was really nothing the Senate could do except release.

The Senate came into emergency session. There were rumors that the President had taken a group of Pentagon folks to California and met with Aliens. Why wasn't the Senate notified? How long had these *Martians* been on our planet? How could they spin this into a political debate? Congress was no better. They knew about the Aliens, hell, half of them had approved Alpha's covert operations and followed the action live on satellite from their living rooms. How the hell did this get out? Pictures from the Presidents meetings with the aliens were being circulated in the Senate and Congress. This was a political crisis. Here were pictures of the President pissing his pants while meeting with an extraterrestrial. Goddamn, she was hot. The Senate spent hours talking about how they could show her a thing or two in the sack. Congress wanted to spin the story into a Women's empowerment message.

Everything went to hell when CNBC reported that the Alien would be holding a press conference in front of City Hall in Los Angeles at noon. Facebook influencers encouraged a giant festival. Every fringe political and religious group demanded a space to protest. Operatives from every County in the world were being dispatched. It was a free-for-all in the heart of the city.

At 11:59 hundreds of thousands of people had gathered in the streets of Los Angeles to participate in the party. Every news organization was represented. All protests were denied, but thousands of signs waived promoting or protesting everything from white supremacy to PETA. The Mayor of Los Angeles was on the steps of City Hall with his administration and riot police were everywhere in attempt to control the crowd.

As the clock struck noon. Two fiery figures landed on the steps of city hall. The intense heat drove the city administrators back into the building. Panic ensued and people were frantically trying to escape the Alien invasion. Vulcán faced the crowd and sent out a wave of comfort that swept over the city center. People stopped pushing and shoving. They helped each other up and brushed each other off. They turned to face the Holy, who now stood on the deserted steps of the City building.

A comet swept across the sky. Television cameras tracked it as it came in, ever brighter, to land between the Holy. More than one news correspondent could be heard to say 'holy shit' on live television. In radiant splendor, the Lord stood in front of the world. The applause of the crowd was deafening, and she stood quietly as it died down. Within minutes, you could hear a pin drop.

Without moving her mouth, Red welcomed all. She sent wave after wave of comfort. These electronic waves carried through the radio and televisions so that all listening felt the warmth and peace that was being communicated.

"We are the children of the Sun. We were exiled from our paradise of fire and heat for offenses to our own kind. We drifted in open space for an eternity. We fell to this blue planet as unintentional guests. In the short time we have been here we have been treated with grace, dignity, and love by good people. Governments have treated us as a weapon or a possession to own. We have come before you today to show that we are here peacefully. That we are blessed to be here and are grateful for the opportunity to share this planet with all creatures. We welcome each of you to get to know us, to share with us, and we with you. We have no secrets and we bring no malice to this beautiful place."

Applause erupted as the truth of these words was draped like a blanket over the crowd. Reinforcing and reaffirming Red's words.

Red continued, "Since we came to this place, we have been hunted by governments and private corporations who wish to use us to further their greed and power. We are not here for that purpose. We will not be the property of any government or corporation. We will not be the tool of any man. And neither should the people of this planet." The crowd went wild. The applause and whistling an affirmation that these people were tired of the manipulation and oppression of the government as well.

"On the Sun, we have a system of justice that maintains the peace and harmony of our Star. At the top are Mer and Zel. They are our Creator's. They bring life and nurture it." A cheer went up for the Creator's.

A shout from the crowd, "are you God?"

The answer was precise and succinct. "We are not gods, nor are the Creator's. They are the mother and father. They are the protectors and the providers. We have no evidence or beliefs that match the Gods of Earth. There is only us, here together, sharing the life experience. Believe what you will, but do not shackle myself, my Holy or the Legion with your Gods."

Questions began to shout out from every direction. Anger rose from those who were devout. Cheers from those who were freethinkers. Red raised her hand, and all calmed immediately.

The Legion are those who are wise and provide counsel and assistance to those in need. I have named some among you as Legion. They wear a cowl of Iron and Cobalt. Seek them out when you have questions or are in need and they will help you.

The Holy are those who have proven themselves to be of the highest character. Those who understand the world and its needs. They are willing to sacrifice their own lives for others. These are my Holy, she said motioning to Vulcán and Kyzyl. There are those among you today who will be asked to join the Holy. When asked, I hope that you will join us.

I am the Lord. I am the judge, the jury, and the executioner of justice. I answer to no government. I appeal to no man. My judgement is rooted in kindness, compassion, and love. If you embrace these things, you will be blessed beyond measure. Those who cannot embrace goodness will be judged.

In the second year after Mongolia, Joel, Tobias, and Tom McGill began a massive mining operation. It was a cavern carved into the granite of the mountain bordering the ranch valley. It was large enough for heavy equipment to drive in and out. They brought out enough gemstones to justify the dig and satisfy the Feds that they were indeed mining, but few if any of the gems came from this location. The cavern got larger with each passing week. When mine inspectors came, the mine was a picture of safety and care. It was built as solid as any transportation tunnel. There were always a few gemstones salted in the tunnel for the inspector to find and pocket. Of course, the electrical needs of a mine this size was significant, and the solar panels soon covered a full twenty acres, feeding into an array of massive Tesla batteries. The mine consumed the power of a small city, compliments of the Sun. Equipment, workers, and deliveries were constant. For months, each was stopped at the bottom of Morris Ranch Road and inspected.

Into the Spring the work continued. Side tunnels came out at different points around the mountain, leaving a spoke-like system consisting of several miles of tunnel. Cattle were brought into graze. A new barn built for the horses. The entire property now resembled a large compound rather than a simple ranch. Life was good.

In the Fall of the year, when the weather cooled and the red-wing blackbirds returned on their way south, Red and her Holy returned to the Ranch. They were greeted by Tobias and Joel. Tom McGill and Grace Washington followed, hand in hand. Tears were shed, strong coffee poured with long, thick dippers of OHO honey. It was a homecoming, and it felt good.

After dinner, Tobias offered that it was time for a surprise. Red look sideways at him. She had read no surprise in his mind. She looked, but nothing was there. He was learning to shield himself she thought with a smile. The humans boarded a few golf carts and Red and the Holy drifted alongside. Red pried and poked at each mind, but she could find no surprise hidden therein. When they arrived at the main entrance of the mine, Red saw the ornately carved concrete entrance. It depicted the story of Reds arrival on Earth and her finding of her Holy. She was truly touched by this gift. Tobias turned on the carts headlights and the convoy drove into the mine for one half mile. They were deep in the mountain when he parked the van and got out. Joel flipped a light switch on the wall and the interior and tunnels lit up like a Christmas tree. Thousands of small, pinpoint lights, creating a beautiful effect in the cavern. The central room was massive, easily fifty yards across. Beautifully carved concrete pillars rose up to meet the ceiling and supported the millions of tons of earth and rock overhead. It was lavishly furnished and comfortable. Its walls were lined with beautiful tourmaline gems, each backlit to produce a colorful effect and each wired together creating walls of powerful energy.

Tobias sighed heavily, struggling to control his emotions. Red floated forward and drifted to the ground inches in front of Tobias. Tears welled up in his eyes and his lip trembled as he fought back the tears. She wrapped warm arms around him, and he broke down, He grabbed her tight and sobbed heavily into her neck. "My girl, my precious girl he said. Welcome home."

"My father," she whispered into his mind, "it is good to be home."

The End

NOTES AND DRAWINGS